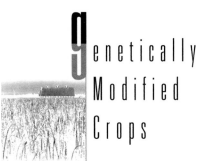

Genetically Modified Crops

Genetically Modified Crops

Nigel G Halford

Crop Performance and Improvement,
Rothamsted Research, UK

Imperial College Press

ICP

Published by

Imperial College Press
57 Shelton Street
Covent Garden
London WC2H 9HE

Distributed by

World Scientific Publishing Co. Pte. Ltd.
5 Toh Tuck Link, Singapore 596224
USA office: Suite 202, 1060 Main Street, River Edge, NJ 07661
UK office: 57 Shelton Street, Covent Garden, London WC2H 9HE

British Library Cataloguing-in-Publication Data
A catalogue record for this book is available from the British Library.

First published 2003
Reprinted 2004

GENETICALLY MODIFIED CROPS

ISBN   1-86094-353-5

Typeset by Stallion Press

Printed in Singapore by World Scientific Printers (S) Pte Ltd

Foreword

Scientists working in plant biotechnology and the associated sciences of molecular and cellular biology and biochemistry in the 1980s and early 1990s looked forward to the first commercial growing of genetically modified (GM) crops with excitement. At last their efforts and expenditure would be justified and, no doubt, more funding would become available for basic and strategic research to underpin a rapidly expanding agricultural biotechnology industry. They would be recognized as being at the forefront of a modern, relevant and beneficial branch of science. As everyone knows, it did not turn out quite like that and as my institute's genetic modification safety officer I became site spokesman on GM crops and was thrust into a debate that was more public and hostile than I had ever expected to be involved in.

This began just before Christmas, 1996, when I was asked to speak to the BBC about a GM story. Soybean and maize from the 1996 harvest, about 2% GM at that time, were being shipped from the US and the GM fraction was not being segregated from the rest. I said yes, expecting a phone call from a researcher later that day. That afternoon I received a call telling me that the BBC had arrived and found a cameraman and reporter waiting for me. I took the crew back over to the lab and had my first interview for television.

Christmas came and the issue went quiet, but Greenpeace, Friends of the Earth and other campaign groups had promoted the GM issue to the top of their list of campaign priorities (Greenpeace gave a February 1997 briefing pack on the issue the title "The End of the World as We Know It") and it was only a matter of time before it came back to the top of the news agenda. It did, thanks to Dr. Pusztai and his predictably sick rats, and the effects of forcing monarch butterfly larvae to eat GM corn pollen. Incidentally, the monarch butterfly prospered after the introduction of GM insect-resistant corn and cotton to large areas of the US in 1996, although unseasonable frosts in its Mexican wintering grounds last year hit it hard. Despite this, I have been assured several times by different people in the United Kingdom that it is extinct and there were supposedly serious reports in New Zealand in 1999 that butterflies could be seen dropping out of the sky across America as they were hit by toxic pollen.

GM crops became a favorite subject not only of the media but of anyone in the United Kingdom who fancied holding a public debate. I have tried to put the science side of the GM debate in meetings from Plymouth to Edinburgh and Pembrokeshire to London, and even Curitiba in Brazil. I have been called lots of rude names and dodged the odd custard pie (non-GM). I have been asked questions on globalization, global warming, CFCs, nuclear power, asbestos, thalidomide, third world debt, Salmonella and, endlessly, BSE, as well as valid topics such as pesticide use, gene flow, antibiotic resistance genes, biodiversity, patenting, labeling, multinationals, and the consequences and opportunities for developing countries.

The outcome of the GM crop debate will have ramifications for science and society that go way beyond plant sciences and agriculture. At present, the world is split in its verdict. While genetic modification is now an established technique in plant breeding throughout many parts of the world, in the United Kingdom and Western Europe its use remains controversial. People have been bombarded by scare stories, misinformation and half truths and have found it hard to obtain answers to their legitimate questions about genetic modification: what is it, how is it done, how does it differ from what has been done before, is it safe, how is it regulated and what implications does it have for plant breeding, agriculture and the environment? The aim of this book is to answer those questions.

CONTENTS

1 DNA, GENES, GENOMES AND PLANT BREEDING

A Brief History of Genetics

To many people the beginnings of genetics can be traced back to the publication in 1859 of Charles Darwin's book, *On the Origin of Species by Means of Natural Selection*. This established the theory of evolution based on the principle of natural selection, discovered independently at the same time by Alfred Russell Wallace, and was the culmination of decades spent collecting and examining evidence.

Darwin argued that species and individuals within species compete with each other. Furthermore, individuals within a species are not all the same, they differ, or show variation. Those that are best fitted for their environments are the most likely to survive, reproduce and pass on their characteristics to the next generation. If the environment changes or a species colonizes a new environment, different characteristics may be selected for, leading to change, or evolution. The diversity of life on Earth could therefore be explained by the adaptation of species and groups of individuals within species to different and changing environments, leading to the extinction of some species and the appearance of others. Species that were similar had arisen from a recent common ancestor. Most controversially, humans were similar to other apes not because God had made it so but because humans and other apes had a relatively recent common ancestor.

1

Darwin was not the only scientist whose thinking was challenging the notion of Life's diversity arising from supernatural creation. Georges Cuvier, in Paris, had noted that fossils in deep rock strata were less like living animals than those in shallow strata. At the same time, studies were revealing that the anatomies of different animals were based on the same internal patterns and Jean-Baptiste Lamarck had already published his theory that animals could transform into one another.

There was a problem with the theory of evolution by natural selection and Darwin was well aware of it. At that time the traits of parents were believed to be mixed in the offspring so that offspring would always be intermediate between their two parents. If this were true, natural selection as Darwin proposed it could not work because it requires there to be variation within a population so that differences can be selected for. Blending traits has the effect of reducing variation with every successive generation. Darwin even considered variations on Lamarck's theory that changes acquired during an organism's lifetime could be passed on to its offspring.

The solution of the problem was found by Gregor Mendel but, ironically, Darwin was never aware of Mendel's work. In 1857, Mendel performed some experiments with pea plants in the garden of his monastery. Mendel noted the different characteristics of the plants, such as height, seed color, seed coat color and pod shape, and observed that offspring sometimes, but not always, showed these same characteristics. In his first experiments, he self-pollinated short and tall plants and found that they bred true, the short having short offspring and the tall having tall offspring. However, when he crossed short and tall plants he found that all of the offspring (the F1 generation) were tall. He crossed the offspring again and the short characteristic reappeared in about a quarter of the next generation (the F2 generation).

Many years and experiments later, Mendel concluded that characteristics were passed from one generation to the next in pairs, one from each parent, and that some characteristics were dominant over others. His experiments concerning the inheritance of a trait in which the seeds were wrinkled are still taught in schools today. His findings were published but ignored for decades as the work of an amateur. Later they became known as the Mendelian Laws and the foundation of modern genetics and plant breeding.

The significance of Mendel's work is that it showed that whether or not the offspring of two parents resemble one parent or are an intermediate between the two, they inherit a single unit of inheritance from each parent. These units are reshuffled in every generation and traits can reappear, so variation is not lost. Units of inheritance subsequently became known as genes.

The next big advance came in 1902, when a British doctor, Sir Archibald Garrod, studied an inherited human disease, Alkaptonuria. Alkaptonuria sufferers excrete dark red urine, because they lack an enzyme that breaks down the reddening agent, alkapton. Garrod noted that Alkaptonuria recurred in families and that parents of sufferers were often closely related to each other. In other words it was an inherited condition and Garrod referred to Alkaptonuria and similar diseases as "inborn errors of metabolism". The significance of Garrod's work was that it made the link between the inheritance of one particular gene and the activity of a single protein.

Like Mendel, Garrod was ahead of his time and his work was forgotten until the link between genes and proteins was made again in the 1930s by American geneticists George Beadle and Edward Tatum. Beadle and Tatum showed that a mutation in the fungus, *Neurospora crassa*, affected the synthesis of a single enzyme required to make an essential nutrient. The mutation was inherited through successive generations. This lead to Beadle and Tatum publishing the one gene, one enzyme hypothesis in 1941. The hypothesis still stands today, although it has been modified slightly to account for the fact that some proteins are made up of more than one subunit and the subunits may be encoded by different genes.

Deoxyribonucleic Acid (DNA)

With the principles of inheritance established, the search was on for the substance through which the instructions for life were passed from one generation to the next. The conclusive experiment that identified this substance is now accepted to be that conducted by Oswald Avery, Colin MacLeod and Maclyn McCarty at the Rockefeller Institute Hospital in 1944. The experiment showed that the transfer of a deoxyribonucleic acid (DNA) molecule from one strain of a bacterium, *Streptomyces pneumoniae*, to another changed its virulence. This showed that DNA was the substance that contained the information for an organism's development, the genetic material. DNA was first discovered in 1869 by Friedrich Miescher, who worked at the Physiological Laboratory of the University of Basel and in Tuebingen. All plants, animals, fungi and bacteria contain DNA and they pass a copy of their DNA to their offspring. The evolution of life on Earth was wholly dependent on the extraordinary nature of DNA and the discovery and characterization of DNA is one of the great achievements of science.

DNA consists of a backbone of molecules of deoxyribose, a type of sugar, linked by phosphate groups (Fig. 1.1). In theory there is no limit to the length of the deoxyribose chain and DNA molecules can be extremely large. Attached to each deoxyribose unit in the chain is an organic base, of which there are four kinds: adenine, cytosine, guanine, and thymine, often represented as A, C, G and T. Any base can be present at any position and it is this that gives DNA its variability. Information is encoded within DNA as the sequence of bases in the chain. In other words, all of the instructions for life on Earth are written in a language of four letters.

The deduction of the three-dimensional structure of DNA is famously attributed to James Watson and Francis Crick, who worked together in Cambridge and published the structure in 1953. In fact, the break-through owed as much to the work of Maurice Wilkins and Rosalind Franklin, two scientists from a team lead by William Bragg who were pioneering the techniques of X-ray crystallography in the study of large molecules in London at that time. Watson and Crick were inspired by the work of Linus Pauling, who had discovered that the molecules of some proteins have helical shapes. They designed several models of DNA and finally came up with the correct structure after analyzing Franklin's X-ray photographs. Watson, Crick and Wilkins jointly received the Nobel Prize for Physiology or Medicine in 1962. Unfortunately, Rosalind Franklin died of cancer in 1958 and could not be honored because the Nobel Prize is not awarded posthumously.

The structure of DNA, as deduced by Watson and Crick, consists of two DNA chains running in opposite directions coiled around each other to form a beautiful structure called a double helix (Fig. 1.2). The bases are on the inside of the helix at right angles to the helix axis and the structure is stabilized by hydrogen bonds between adjacent bases from each chain. Because of the dimensions of the helix and the structures of the bases, adenine on one chain must always be faced with thymine on the other, while guanine is always paired with cytosine.

This specific pairing of the bases is important because it underlies the process by which DNA can be replicated as an exact copy. If double-stranded DNA is unravelled to form two single strands, each strand can

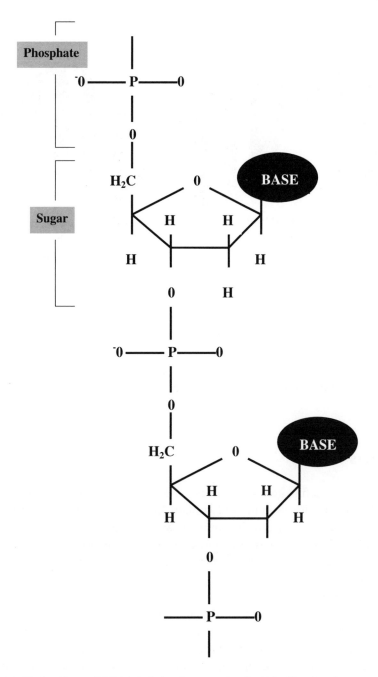

Fig. 1.1. The backbone of DNA is a chain of sugar molecules linked by phosphate groups.

**Sugar phosphate
backbones**

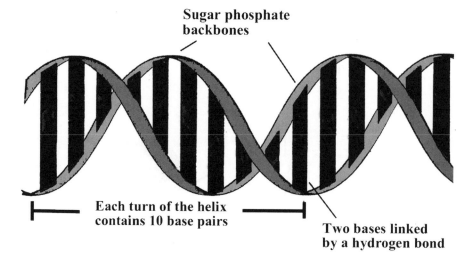

**Each turn of the helix
contains 10 base pairs**

**Two bases linked
by a hydrogen bond**

Fig. 1.2. Structure of a DNA double helix.

act as a template for the synthesis of a complimentary chain and two
replicas of the original double-stranded molecule are created.

Genes

We can now define genes, the units of heredity, as functional units within
a DNA molecule. A gene will contain the information not only for the
structure of a protein, but also for when and where in an organism the gene
is active. This information is encoded in the sequence of the base pairs
within the region of the DNA molecule that makes up the gene. Genes can
comprise over a million base pairs but are usually much smaller, averaging
three thousand base pairs.

The part of the gene containing the information for the primary structure of the protein encoded by the gene is called the coding region.
The sequence of amino acids in the protein is determined by the sequence
of base pairs in the coding region, each amino acid in the protein being
represented by a triplet of base pairs in the DNA sequence. Adjacent to

(or "up-" and "down-stream" of) the coding sequence are the regulatory regions of the gene that determine when and where the gene is active (or expressed). Most of this regulatory information is usually, but not always, "upstream" of the coding region in what is called the gene promoter (Fig. 1.3). The region "downstream" of the coding region is called the gene terminator.

Gene Expression

The term gene expression refers to the process by which the protein that is encoded by the gene is produced. It comprises the following sequence of events:

(1) The DNA molecule is used as a template for the synthesis of a related, single-stranded molecule called ribonucleic acid (RNA). RNA, like DNA, comprises organic bases on a sugar-phosphate backbone. The sequence of bases on the newly synthesized RNA molecule is determined by the sequence of bases on the DNA template, hence the encoded information is passed from the DNA to the RNA molecule. This process is called transcription.

(2) The RNA molecule is processed and transported to protein complexes called ribosomes where protein synthesis occurs.

(3) A protein is synthesized, the amino acid sequence of which is specified by the RNA molecule. This process is called translation. Information encoded in the DNA molecule has been transferred through an RNA molecule to the protein production machinery and used to make a protein.

Genes that are active (i.e., being expressed) throughout an organism all of the time are referred to as constitutive or house-keeping genes. Most genes are not constitutive, but are subject to various kinds of regulation. Some are expressed only in certain organs, tissues or cell types. Others are expressed during specific developmental stages of an organism, or are expressed in response to a stimulus. In the case of plants, gene expression responds to a host of stimuli, including light, temperature, frost, grazing, disease, shading and nutritional status. DNA is not just the blue-print for an organism's development; it is

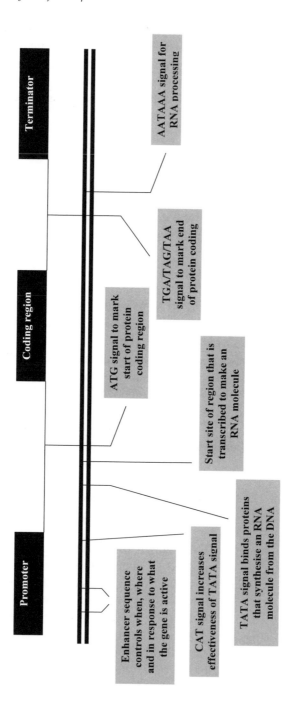

Fig. 1.3. Schematic diagram of the structure of a gene.

active throughout an organism's life and plays a central role in an organism's interaction with its environment.

Genomes

The genome is defined as an organism's complete compliment of DNA. Genomes can be very large but it is now possible to determine the base pair sequences of entire genomes from even complex animals, plants and fungi. The entire yeast genome sequence was completed in 1996 and a draft of the human genome sequence was completed in 2000. The sequences of two entire plant genomes have also been determined. These are arabidopsis (a plant also known as thale cress that is used as a model in plant genetics because of its relatively small genome) and rice.

The human genome contains approximately three billion base pairs, organized into 23 chromosomes ranging from 50–250 million base pairs long. The rice genome contains 466 million base pairs, while that of arabidopsis contains approximately 126 million base pairs. The human genome, however, contains only 30 000–40 000 genes, while that of rice contains 45 000–56 000 genes (one could argue that rice is a more complex and highly evolved organism than man).

The reason for the discrepancy between genome size and gene number is that much of the genome does not contain genes. In fact, genes represent only 2% of the human genome. The rest of the DNA is sometimes referred to as "junk DNA" and the amount of it varies greatly between different species. Much of it is highly repetitive in its base pair sequence and it may promote rearrangement of the DNA molecules, driving the formation of new genes and therefore variation and evolution.

Genetic Change

Genetic change is a natural and desirable process. It results in variation in shape, form and behavior of the individuals within a species, allowing for evolution and adaptation. It is crucial for the survival of any species and the evolution of new species in response to environmental change.

The process of genetic change in nature is called evolution and when driven by natural selection it results in adaptive improvement. In the absence of selection it leads to diversity and variation, making it more likely that some individuals will survive if the environment changes.

Plant Breeding

Natural selection is not the only driver for genetic change at work in the world today. Man has discovered that rapid genetic change can be induced by the artificial selection of individuals for breeding. This has been used to produce the Chihuahua, the greyhound and the bull terrier from the wolf, thoroughbred racehorses, the Shire horse and the Shetland pony, and every modern farm animal. It has also been applied for millennia to crop improvement.

It is probable that crop improvement has been practiced since humankind first started to plant and harvest crops, rather than forage for food from wild plants. This is thought to have become widespread about 10 000 years ago. At first, such improvement may well have occurred unconsciously, by harvesting and growing on the most vigorous individuals from highly variable populations, but then became more systematic. Types of a crop plant with different characteristics would be grown in adjacent plots and some of the seed produced would result from crossing of the two types. Farmers would then select the best seed for the next generation. This relatively primitive but effective form of plant breeding is still used in many parts of the world today and through the ages has changed crop plants greatly from their wild ancestors and relatives.

This is illustrated quite dramatically in Fig. 1.4, in which grain from a modern breadmaking wheat variety is compared with grain from two wild relatives. Wheat cultivation and improvement enabled the production of food in large quantities and it is doubtful whether modern civilization would have developed without it. The development of modern wheat is a good example of how the genetics of crop plants have been manipulated by farmers and breeders who for millennia knew nothing about the molecular basis of what they were doing.

Wheat actually comprises many different species, some with one, two or even three genomes (diploids, tetraploids and hexaploids, respectively). The tetraploids, which include cultivated pasta wheats, arose through hybridization and genome duplication between ancestral diploid species. Breadmaking wheat is a hexaploid that arose through hybridization between a tetraploid and a third diploid species. There are no wild

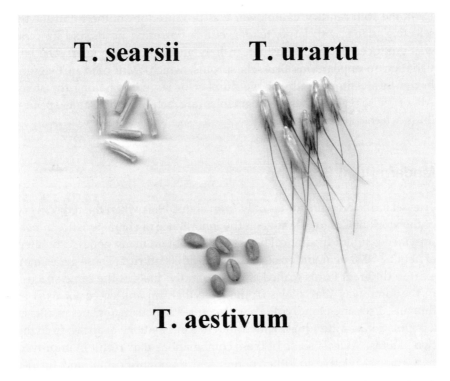

Fig. 1.4. Bread wheat, *Triticum aestivum*, is a product of thousands of years of selective breeding. Here its grain is compared with that of two wild relatives, *Triticum searsii* and *Triticum urartu*.

hexaploid wheat species. Breadmaking wheat appeared within cultivation in southwest Asia approximately 10 000 years ago and its use spread westwards into Europe.

   Another crop family, the cabbage family of vegetables, which includes kale, cabbage, cauliflower, broccoli and Brussels sprouts, shows what remarkable changes have been introduced into crop plants by simple selection over many generations. The wild relative of the cabbage family grows in the Mediterranean region of Europe, and it was first domesticated approximately 7000 years ago. Through selective breeding, the crop plants became larger and leafier, until a plant very similar to modern kale was produced in the 5th century BC. By the first century AD, a different variation was being grown alongside kale. It had a cluster of tender young leaves at the top of the plant and is known today as cabbage.

In the 15th century, cauliflower was produced in southern Europe by selecting plants with large, tender, edible flowering heads and broccoli was produced in similar fashion in Italy about a century later. The last variation to appear was Brussels sprouts, which were bred in Belgium in the 18th century, with a large number of large buds along the stem. All of these very different vegetables are actually the same species, *Brassica oleracea.*

Modern Plant Breeding

True scientific breeding dates only from about 1900, when the rediscovery of the work of Gregor Mendel on the inheritance of characteristics in pea, provided a sound theoretical basis. Plants contain many genes, estimated at about 26 000 in arabidopsis and 45 000–56 000 in rice. These genes may exist in different forms (called alleles) in individuals of the same species, in the same way that alleles of the gene that imparts eye color exist in humans. Crossing of individuals with contrasting characteristics results in a population of individuals with different combinations of alleles from the two parents. At least some of these combinations may result in improved performance relative to either parent and the identification and further improvement of these is the task of the plant breeder. Since the breeder is literally dealing with thousands of genes, the task is formidable.

Nevertheless, plant breeders have been incredibly successful at improving crop yield and it is just as well that they have. At the end of the 18th century, Reverend Thomas Malthus predicted in his *Essay on the Principle of Population*, which he published anonymously, that food supply could not keep up with exponentially rising population growth. At the time, world population was approximately one billion. In 1999, the world population reached six billion, and yet, with the exception of sub-Saharan Africa, that population has generally not been critically short of food.

This has been achieved through dramatic increases in crop yield. As an example, the yield per hectare of wheat grown in the United Kingdom over the last 800 years is shown in Fig. 1.5. It has increased approximately ten-fold, with more than half of that increase coming since 1900. Of course, plant breeding has not been the only contributor to this increase. Mechanization, the development and use of nitrogen fertilizers, herbicides and pesticides and other improvements in farm practice have played a part. Indeed, it was the combination of mechanization, the

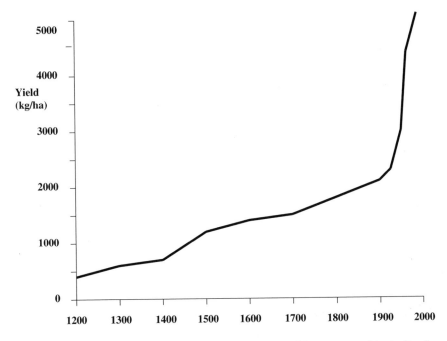

Fig. 1.5. UK wheat yield, 1200 to present day. (Data from US Department of Agriculture)

widespread use of agrochemical and the incorporation of dwarfing genes into cereals that led to the "Green Revolution" of the 1960s and 1970s. The dwarfing genes concerned actually affected the synthesis of a plant hormone, gibberellin, although that was not known at the time. Their incorporation reduced the amount of resources that crop plants put into their inedible parts and at the same time made them less susceptible to lodging (literally falling over in damp and/or windy conditions).

Plant breeding has also been successful in improving food safety. Contrary to the popular notion that natural is "good" and man-made is "bad" when it comes to food, plants did not evolve to be eaten. In fact the opposite is true, and since plants are unable to flee a large animal or brush off an insect and do not have an immune system to fight off micro-organisms, they have evolved a sophisticated armoury of chemical weapons with which to defend themselves. Many of these chemicals are potentially toxic or allergenic to humans and they are particularly abundant in seeds and tubers, whose rich reserves of proteins, starch and oil are particularly attractive to herbivores, pests and pathogens. Well-known examples are glycoalkaloids in potatoes, cyanogenic glycosides in linseed,

proteinase inhibitors in soybean and other legume seeds and glucosinolates in *Brassica* oilseeds.

Plant breeders have made great strides in reducing the levels of these
chemicals in crop plants, although some crops have only been regarded
as fit for human consumption surprisingly recently. Oilseed rape, for
example, was first grown in the UK during World War II to provide oil
for industrial uses and some varieties are still grown for that purpose.
Its oil was regarded as unfit for human consumption because it contained
high levels of erucic acid and glucosinolates. Glucosinolates were present
at even higher levels in the meal. They have a bitter/hot flavor, are
very poisonous and probably explain why oilseed rape has few specialized pests. After World War II, breeders gradually reduced the levels of
erucic acid and glucosinolates and the oil was finally passed for human
consumption and the meal for animal feed in the 1980s.

Wide and Forced Crossing and Embryo Rescue

There is a major limitation to the success of plant breeding, which is the
extent of variation in the parental lines. Plant breeders cannot select for
variation that is not present in their breeding population. They have
therefore looked further afield for genes to confer important characters,
including exotic varieties and wild relatives of crop species. This "wide
crossing" would not occur in nature and usually requires rescue of the
embryo to prevent abortion. This entails surface sterilizing developing
seeds and dissecting them under a microscope to remove the embryo. The
embryos are then cultured in a nutrient medium until they germinate and
develop into seedlings. These are then transferred to soil.

Forced crosses of this sort may be carried out to transfer genes for
single characters (disease resistance, for example). However, tens of thousands of other genes will inevitably be brought into the breeding program
alongside the desired gene. Most of these can be eliminated by crossing the
hybrid plant repeatedly with the crop parent (backcrossing). Nevertheless,
many genes of unknown function will remain at the end of this process
and if they have undesirable effects on the performance of the crop the
program will fail.

Forced crosses have even been made between different plant species.
The best-known example of this is triticale, a hybrid between wheat

and rye. The first deliberately-made hybrid between wheat and rye was reported by A. S. Wilson in Scotland in 1875. However, these plants were sterile because the chromosomes from wheat and rye will not form pairs. This can be overcome by inducing chromosome doubling by chemical treatment, usually with colchicine. The cross is usually made between durum wheat, a tetraploid, and rye to produce a hexaploid triticale, although it is also possible to cross hexaploid wheat with rye to produce an octoploid triticale.

The name triticale was used first in 1935 by Tschermak but it was not until 1969, after considerable improvement through breeding, that the first commercial varieties of triticale were released. Triticale is now grown on more than 2.4 million hectares worldwide, producing more than six million tonnes of grain per year. Its advantage is that it possesses the yield potential of wheat and the hardiness (tolerance of acid soils, damp conditions and extreme temperatures) of rye. However, it does not match the breadmaking quality of wheat and is used mostly for animal feed.

Radiation and Chemical Mutagenesis

Another way of increasing the variation within a breeding population is to introduce mutations artificially. This is usually done with seeds and involves treatment with neutrons, gamma rays, X-rays, UV radiation or a chemical (a mutagen). All of these treatments damage DNA. If there is too much damage the seed will die, but minor damage can be repaired, resulting in changes in the DNA sequence. Sometimes even small changes of this sort are lethal, but occasionally changes are made to a gene that leave the seed viable but alter the characteristics of the plant. The process is entirely random and mutagenesis programs usually involve very large populations of at least 10 000 individuals.

The first attempts to produce plant mutants were made in the 1920s and the first commercial varieties arising from mutation breeding programs became available in the 1950s. The technique was particularly fashionable in the 1960s and 1970s but continues to be used today. One of the earliest cultivars produced by mutagenesis was the oilseed rape cultivar Regina II, which was released in Canada in 1953. Mutagenesis played an important role in the improvement of oil quality of oilseed rape and flax, and in durum wheat breeding in Italy. There are estimated to be 200 rice varieties that incorporate artificially induced mutations and most

North American white bean varieties incorporate a mutation that was first induced by X-ray treatment.

The best known example of an irradiation mutant grown in the UK is the barley variety, Golden Promise. This was the most successful UK malting barley variety in the latter part of the last century and although it is not grown widely now it has been incorporated into many barley breeding programs.

The Advent of Genetic Modification

Crossing different varieties and species involves the mixing of tens of thousands of genes, sometimes with unpredictable results. Mutagenesis is, of course, a random process and it is impossible to count, let alone characterize, all of the effects of a mutagenesis program. Radiation and chemical mutants are likely to carry a baggage of uncharacterized genetic changes with unknown effects. Nevertheless, these techniques were developed and applied to plant breeding at a time when food production was a much higher priority than food safety and the risks were not a matter of public debate. They have now been around for so long that they are considered to be part of "traditional" plant breeding and crops modified by wide crossing and mutation breeding are accepted readily.

Even with the possibility of crossing crop plants with exotic relatives and related species and introducing mutations artificially, the variation available to plant breeders remained limited. Both techniques also had the disadvantage of introducing unwanted as well as desirable genetic changes. In the late 1970s, a new technique became available, that of inserting specific genes into the genome of a plant artificially. It meant that technically there was no limit to the source of new genes and it enabled plant breeders to bring specific genes into breeding programs without unwanted genetic baggage. This new technique was first called genetic engineering and subsequently genetic modification.

2 THE TECHNIQUES OF PLANT GENETIC MODIFICATION

Any new crop variety has been changed genetically, using the methods described in the previous chapter. However, the term genetically modified (GM) is a relatively new expression that describes a plant that contains a gene or genes that have been introduced artificially. Such plants are also described as being transgenic or having been transformed. The term genetically engineered (GE) can be used instead of genetically modified. Plant genetic modification became possible in the late 1970s as a result of the development of techniques for manipulating DNA in the laboratory and introducing it into the DNA of a plant.

A Brief History of the Development of Recombinant DNA Technology

In the quarter century after the description of the structure of DNA by Watson and Crick in 1953, huge strides were made in the study of DNA and the enzymes present in cells that can work on it. In 1955, Arthur Kornberg at Stanford University isolated DNA polymerase, an enzyme that synthesizes DNA. He received the Nobel Prize for Medicine in 1959. In 1966, Bernard Weiss and Charles Richardson working at Johns Hopkins University isolated DNA ligase, an enzyme that glues two ends of DNA together. In 1970, Hamilton Smith, also at Johns

17

Hopkins University, described the first characterization of a restriction endonuclease (now usually called a restriction enzyme), an enzyme that had the ability to recognize specific short sequences of base pairs in a DNA molecule and cut the molecule at that point. Smith shared the Nobel Prize for Medicine in 1978.

The rapid progress continued. In 1972, Paul Berg at Stanford reported that he had constructed a DNA molecule by cutting viral and bacterial DNA sequences with restriction enzymes and then recombining them. He received the Nobel Prize for Chemistry in 1980. Then in 1973, Stanley Cohen and Annie Chang of Stanford together with Herbert Boyer and Robert Helling of the University of California, San Francisco, demonstrated that DNA that had been cut with a restriction enzyme could be recombined with small self-replicating DNA molecules from bacteria called plasmids. The new plasmid could then be reintroduced into bacterial cells and would replicate. In 1977, Walter Gilbert at Harvard and Fred Sanger in Cambridge separately developed methods for determining the sequence of base pairs in a DNA molecule. Sanger and Gilbert shared the Nobel Prize for Chemistry with Berg in 1980.

Scientists now had the tools for cutting DNA molecules at specific points and gluing them back together in different combinations to make new molecules. This was known as recombinant DNA technology. It meant that DNA from any source could be cloned and bulked up in bacteria for analysis. The bacterium of choice for this purpose is *Escherichia coli*. This is a human gut bacterium, although the strains used in the laboratory have been modified so that they are not pathogenic. The cutting and splicing of plant, animal and fungal genes into plasmids and their propagation in *Escherichia coli* underpinned the molecular analysis of gene structure and function. It also lead to the first commercial use of recombinant DNA technology in the pharmaceutical industry, recombinant human insulin produced in *Escherichia coli* being approved by the Food and Drug Administration of the US in 1981.

Agrobacterium Tumefaciens

Another bacterial species that contains a plasmid is the common soil bacterium, *Agrobacterium tumefaciens*, a bacterium that infects wounded plant tissue and causes the disease known as crown gall. In 1977, Nester, Gordon and Dell-Chilton showed that genes from a plasmid carried by

Agrobacterium tumefaciens were inserted into the DNA of host plant cells. Effectively, the bacterium could modify host plant cells genetically. The plasmid was called the tumor-inducing or Ti plasmid. This observation lead to *Agrobacterium tumefaciens* becoming one of the most reliable and widely-used means of transferring foreign DNA into plants.

During infection, the host plant cells release phenolic compounds. These compounds interact with proteins encoded by virulence (VIR) genes carried by the Ti plasmid and induce the bacteria to bind to plant cell walls. Other VIR genes on the Ti plasmid then become active and they cause a single strand of DNA to be nicked out of a region of the Ti plasmid. This T-DNA (transfer DNA) is protected by specialized proteins and is transferred to the adjacent plant cell where it integrates into the plant DNA.

Once integrated into the plant DNA, genes present in the T-DNA become active. These genes encode proteins that perturb the normal hormone balance of the cell. The cell begins to grow and divide to form a tumor-like growth called the crown gall. The cells of the crown gall are not differentiated, in other words they do not develop into the specialized cells of a normal plant. They can be removed from the plant and cultured as long as they are supplied with light and nutrients and are protected from fungal and bacterial infection. A clump of these undifferentiated cells is called a callus.

All of the cells of a crown gall or a callus contain the T-DNA that was inserted into the original host cell. Indeed, the T-DNA will be inherited stably through callus culture and through generations of GM plants that are produced from the callus. The T-DNA contains another set of genes that induce the host cell to make and secrete unusual sugar and amino acid derivatives that are called opines, on which the *Agrobacterium* feeds. There are several types of opine, including nopaline and octopine, and different types are produced after infection with different strains of the bacterium. The genes that enable the bacterium to feed on the opines are also carried by the Ti plasmid.

Agrobacterium tumefaciens has a close relative, *Agrobacterium rhizogenes*, that causes hairy root disease. *Agrobacterium rhizogenes* uses a similar strategy to *Agrobacterium tumefaciens*, and carries an Ri plasmid that functions in a similar way to the Ti plasmid but induces a different response in the host plant cell. Together they are nature's genetic engineers and were using plant genetic modification millions of years before humans "invented" it.

Use of *Agrobacterium tumefaciens* in Plant Genetic Modification

The natural mechanism of plant genetic modification by *Agrobacterium tumefaciens* can be used to introduce any gene into a plant cell. This is because the only parts of the T-DNA that are required for the transfer process are short regions of 25 base pairs at each end. Anything between these border regions will be transferred into the DNA of the host plant cell. This has allowed the development of plasmids that contain the left and right borders of the T-DNA, but none of the genes present in "wild-type" T-DNA. The most widely used of these plasmids are binary vectors, so-called because they will replicate in either *Agrobacterium tumefaciens* or *Escherichia coli*. This means that the plasmid can be manipulated to insert new genes between the T-DNA borders and then bulked up in *Escherichia coli* before transfer to *Agrobacterium tumefaciens*. Binary vectors do not contain the *VIR* genes present on the Ti plasmid, so they are unable to induce transfer of the T-DNA into a plant cell on their own. However, when the binary vector is present in *Agrobacterium tumefaciens* together with another plasmid containing the *VIR* genes, the region of DNA between the T-DNA borders, including the genes that have been placed there in the laboratory, will be inserted into the DNA of a host plant cell.

The most widely used method for producing GM plants using this system is to inoculate sterile explants (for example leaf pieces, stem sections or tuber discs) with *Agrobacterium tumefaciens* containing a binary vector and a Ti plasmid lacking its own T-DNA. The explants are transferred to sterile petri dishes containing solid medium that induces the cells that have received the T-DNA to form callus (Fig. 2.1a). Calli are then transferred onto a second medium that contains a plant hormone that induces the callus to form a shoot (Fig. 2.1b). Once a shoot with a stem has formed, it is transferred to a third medium that does not contain the shoot-inducing hormone. Hormones produced by the shoot itself then induce root formation and a complete plantlet is formed (Fig. 2.1c). Up to this point the process has to be done in sterile conditions to prevent bacterial or fungal infection. Once shoot and root are fully formed, the plantlet can be transferred to soil and treated like any other plant (Fig. 2.1d).

All of the cells of the plant will contain the T-DNA integrated into its own DNA and the T-DNA and all of the genes in it will be inherited in the same way as the other genes of the plant.

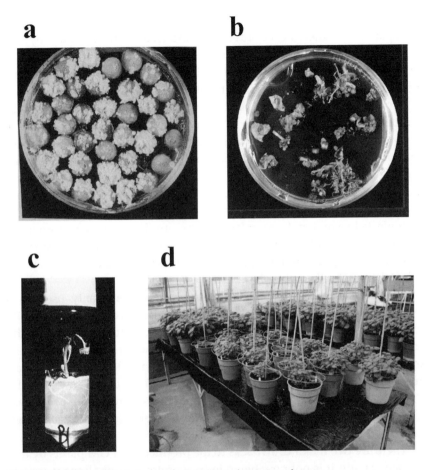

Fig. 2.1. *Agrobacterium tumefaciens*-mediated transformation of potato.

a. Potato tuber discs that have been infected with *Agrobacterium tumefaciens*, showing the formation of clumps of undifferentiated cells called callus. Cells within the callus contain genes that have been inserted into the potato genome by the bacterium.

b. Shoots induced from callus by the application of a plant hormone.

c. A complete GM potato plantlet.

d. Once the plantlets are large enough they can be transferred to soil in pots and grown in containment in a greenhouse.

Transformation of Protoplasts

Protoplasts are plant cells without a cell wall. The cells are usually derived from leaf tissue and are incubated with enzymes (cellulases, pectinases and hemicellulases) that digest away the cell wall.

Agrobacterium tumefaciens will infect and transform protoplasts. The protoplasts can then be cultured and induced to form callus by the application of plant hormones (auxins and cytokinins). GM plants can be regenerated from the callus as described for *Agrobacterium tumefaciens*-mediated transformation of explant material.

Protoplasts can also be induced to take up DNA directly. This process is called direct gene transfer or DNA-mediated gene transfer. There are two widely used methods for achieving this. The first involves treatment with polyethylene glycol (PEG) or a similar polyvalent cation. The exact way in which PEG works is not known, but it is believed to act by causing the DNA to come out of solution and by eliminating charge repulsion. The second method is called electroporation and involves subjecting the protoplasts to a high voltage pulse of electricity. This causes the formation of pores in the plant cell membrane and although these must be repaired very rapidly to prevent the protoplasts from dying, it briefly allows DNA to be taken up by the protoplasts.

Direct gene transfer into protoplasts is most commonly used as a research tool. A gene of interest is introduced into protoplasts so that its activity and function can be studied. The gene does not integrate into the protoplasts' own DNA and is eventually broken down, so the protoplasts are only temporarily, or transiently, transformed. However, in a small proportion of the protoplasts, the introduced DNA will integrate into the host DNA and the protoplast will be stably transformed. The protoplast can then be induced to form callus and a GM plant can be regenerated from it.

There are two major drawbacks to the production of GM plants by this method. Firstly, it is not possible to induce protoplasts of all plant species to form callus and regenerate a whole plant. Secondly, the introduced DNA is often rearranged and does not function as expected.

Particle Gun

Flowering plants can be subdivided into two subclasses, monocotyledonous and dicotyledonous, depending on their embryo structure. In the wild, *Agrobacterium tumefaciens* only infects dicotyledonous plants and although progress has since been made in adapting its use for the genetic modification of monocotyledonous plants in the laboratory, and it is now the method of choice for some monocotyledonous species, its use was

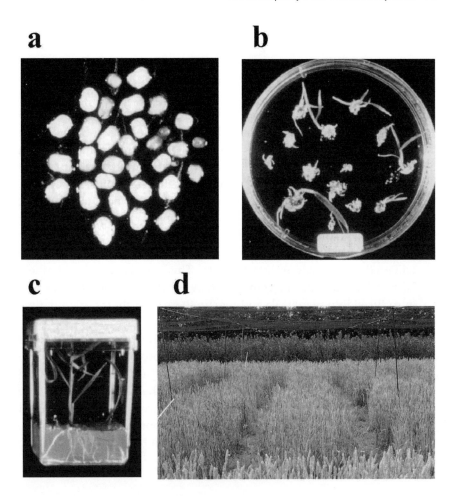

Fig. 2.2. Transformation of wheat by particle bombardment.

a. Wheat embryos are the targets for bombardment with tiny gold particles coated with DNA. Some of the cells within the embryos take up the DNA and incorporate it within their own DNA.

b. The genetically modified cells are induced to form callus material and then shoots by the addition of plant hormones.

c. Shoots are placed on a different medium lacking hormones and begin to form roots.

d. Once whole plants have been produced they can be treated in the same way as non-GM wheat plants. This picture shows a field-trial of GM wheat at Long Ashton Research Station near Bristol, UK.

Pictures kindly provided by Pilar Barcelo (a–c) and Peter Shewry (d).

limited at first to dicotyledonous plants. This was important because cereals, including major crop species such as wheat, maize and rice, belong to the monocotyledonous subclass and these species are also not amenable to the regeneration of whole plants from protoplasts.

Transformation of cereals eventually became possible with the invention of the particle bombardment method. In this method, plant cells are bombarded with tiny particles coated with DNA. Some of the DNA is washed off the particles and becomes integrated into the plant genome. This is carried out in a particle gun, the first of which used a small explosive charge to bombard the plant cells with tungsten particles. Subsequently, a variety of devices were developed but one using a burst of pressurized helium gas in place of the explosive charge has been the most successful and gold particles are now used instead of tungsten. There are now hand-held versions of this device on the market, allowing DNA to be delivered to cells within intact plants. This has become an extremely useful tool for studying the transient activity and function of genes that are introduced into plant cells and remain there for a short period of time but do not integrate into the host plant DNA.

Particle bombardment, which has acquired the unfortunate name of biolistics, has been particularly successful in the production of genetically modified cereals, including maize, wheat, barley, rice, rye and oat and the first commercial GM cereal crops (Chapter 3). The plant tissues that are used are either isolated explants (Fig. 2.2a) that are bombarded, induced to become embryogenic and regenerated, or embryogenic cell cultures. Whole plants are then regenerated from single embryonic cells that have taken up the DNA (Figs. 2.2b–d).

Other Direct Gene Transfer Methods

The technique of electroporation can be applied to intact cells in tissue pieces or in suspension, as well as to protoplasts, but has only been shown to work efficiently in a few species. The other direct gene transfer method to have been developed is silicon carbide fiber vortexing. Plant cells are suspended in a medium containing DNA and microscopic silicon carbide fibers. The suspension is vortexed and the fibers penetrate the plant cell walls, allowing the DNA to enter.

Agrobacterium-Mediated Transformation without Tissue Culture

The methods of plant transformation through the infection of explants or protoplasts with *Agrobacterium tumefaciens* and the regeneration of intact plants is relatively straightforward with many plant species, including major crops such as soybean and potato. It does have the drawback, however, along with all of the direct gene transfer methods, that the regeneration of plants from single cells often gives rise to mutation. Plants carrying harmful mutations have to be screened out before the GM plant is incorporated into a breeding program. If the genetic modification is being made in order to investigate the function of a particular gene for research purposes, the effects of the genetic modification may be difficult to distinguish from the effects of random mutations.

For this reason, and in order to speed up the process, methods of plant transformation that do not require tissue culture have been developed. The most successful of these is floral dip transformation. This was developed by Georges Pelletier, Nicole Bechtold and Jeff Ellis using the model plant for genetic research, arabidopsis. Plants at the early stages of flowering are placed in a suspension of *Agrobacterium tumefaciens* in a vacuum jar, a vacuum is applied to remove air surrounding the plant tissue and allow the bacteria to come into contact with it, the plants are grown to seed and approximately 1% of the seeds will be genetically modified. This method is now widely used in basic research using arabidopsis and has been adapted with some success for use with other plant species, including soybean and rice.

Selectable Marker Genes

A limitation to all of the plant genetic modification techniques described above is that only some of the cells in the target tissue are genetically modified, irrespective of the method of DNA transfer. It is, therefore, necessary to kill all of the cells or regenerating plants that are not modified and this requires that the presence of the gene of interest can be selected for or that the gene of interest be accompanied by at least one other gene that acts as a selectable marker. The regeneration of GM plants from a transformed plant cell is carried out in the presence of a selective agent, tolerance of which is imparted by the marker gene.

In practice, selectable marker genes make the transformed cells and GM plant resistant to an antibiotic (Fig. 2.3a) or tolerant of a herbicide (Fig. 2.3b). The safety of antibiotic resistance genes in plant biotechnology is discussed in Chapter 5. Those that are used confer resistance to kanamycin, geneticin or paromycin (collectively known as aminoglycosides) or to hygromycin, none of which have critical, if any use in medicine. Kanamycin, geneticin and paromycin resistance is imparted by a gene called *NPTII* (neomycin phosphotransferase) while hygromycin resistance is imparted by one of *HPT*, *HPH* or *APH-IV* (hygromycin phosphotransferase). Antibiotic resistance genes are widespread in nature (Chapter 5), and all of these genes were obtained originally from the common gut bacterium, *Escherichia coli*. The *NPTII* gene is probably the most widely used selective marker gene in the genetic modification of dicotyledonous plants. However, it is not generally used in the genetic modification of cereals because kanamycin is not toxic enough to cereal cells and plants.

GM crops with herbicide tolerance are now widely used commercially (Chapter 3). Obviously, genes that impart tolerance of a herbicide do not need to be accompanied by an additional selective marker. GM plants are simply regenerated in the presence of the herbicide to prevent non-GM plants from growing. The use of herbicide tolerance genes as selectable markers to accompany other genes of interest was developed because the removal of non-GM cells and regenerating plants with antibiotics was found to be unreliable with cereals. They work either by encoding a modified or alternative, insensitive version of a protein that is targeted by the herbicide, or by encoding an enzyme that breaks the herbicide down.

The most widely used herbicide tolerance genes for research purposes are the *BAR* or *PAT* (phosphinothricin acetyl transferase) genes, from the bacteria *Streptomyces hygroscopicus* and *Streptomyces viridochromogenes*. They impart tolerance to herbicides based on phosphinothricin, including gluphosinate and bialaphos. They are also used in some commercial herbicide-tolerant GM crop varieties (Chapter 3). This selection method works for all of the cereals that have been transformed to date. Less widely used for research but also with applications in commercial GM crop varieties are genes that impart tolerance of glyphosate, such as the *EPSPS* (enolpyruvylshikimate phosphate synthase) and *GOX* (glyphosate oxidoreductase) genes from *Agrobacterium*. Finally, there are genes that impart tolerance of sulfonylurea compounds, such as the *ALS* (acetolactate synthase) gene from maize.

a

b

Fig. 2.3. Antibiotic resistance and herbicide tolerance marker genes.

Plants that have been genetically modified have to be separated from plants that have not. If the gene that has been engineered into the plant cannot be selected for, it must be accompanied by a marker gene that can. The most common marker genes impart resistance to an antibiotic or tolerance of a herbicide.

a. A genetically modified tobacco plant (right) carrying a gene that imparts resistance to an antibiotic, kanamycin, and an unmodified plant (left) growing in the presence of the antibiotic.
b. A genetically modified wheat plant (right) carrying a gene that imparts tolerance of a herbicide, bialaphos, and an unmodified plant (left) growing in the presence of the herbicide.

Picture kindly provided by Pilar Barcelo.

a

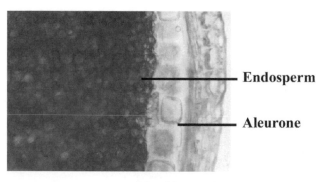

Endosperm

Aleurone

|---- 0.1 mm ----|

b

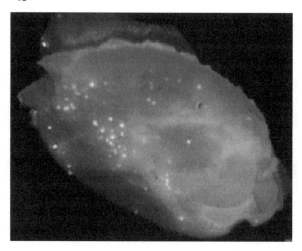

Fig. 2.4. *UidA* (Gus) and Green Fluorescent Protein (GFP) visible marker genes.

a. The *UidA* gene is bacterial in origin and encodes an enzyme called β-glucuronidase. When supplied with the appropriate substrate this enzyme will produce a blue product, allowing the exact location of gene activity to be pinpointed. In this picture the gene is under the control of a wheat gene promoter that is active only in the seed endosperm (the major seed storage tissue). The picture contrasts the dark-stained endosperm cells with the unstained cells of the surrounding tissue called the aleurone.

b. The product of the GFP gene (from the jellyfish *Aequorea victoria*) fluoresces when excited at 400 nm and is visualized using a fluorescence stereomicroscope. This picture shows bright fluorescent foci in a wheat embryo that has been bombarded with a GFP gene controlled by a wheat promoter that is active in this tissue.
Picture kindly provided by Sophie Laurie.

Visual/Scoreable Marker Genes

A marker gene that encodes a visual or scoreable product may be used alongside a selectable marker gene in order to allow GM cells to be visualized. Examples of genes used in this way are a bacterial gene, *UidA* (commonly called Gus), a jellyfish green fluorescent protein (GFP) gene and a luciferase (Lux) gene. The *UidA* gene encodes an enzyme called β-glucuronidase. When supplied with the appropriate substrate this enzyme will produce a product that fluoresces under ultra-violet light. The intensity of fluorescence can be measured using a fluorimeter, giving an indication of the amount of product and therefore the amount of enzyme that is being produced and the activity of the gene. Plant tissue is ground up in this experiment to allow the enzyme and substrate to come together. An alternative substrate can be used to penetrate thin sections of plant tissue. The β-glucuronidase enzyme produces a blue product from this substrate and the blue color can be seen when the tissue is examined through a microscope, allowing the exact location of gene activity to be pinpointed (Fig. 2.4a).

The *UidA* gene has proved extremely valuable as a research tool. Its main drawback is that it is destructive; the plant tissue is killed by the assay. The products of the luciferase and GFP genes, on the other hand, can be visualized without killing the plant tissue and have little or no toxicity for the plant cell. The favored luciferase system actually requires two genes, *LuxA* and *LuxB*, from a bacterium, *Vibrio harveyi*. Activity of the two genes and the combining of the two gene products results in bio-luminescence within the tissue and this can be detected and measured by light-sensitive equipment.

The GFP gene comes from an intensely luminescent jellyfish, *Aequorea victoria*. Light-emitting granules are present in clusters of cells around the margin of the jellyfish umbrella and contain two proteins, aequorin, which emits blue-green light, and green fluorescent protein (GFP), which accepts energy from aequorin and re-emits it as green light. GFP fluoresces maximally when excited at 400 nm and it is visualized in GM plants using a fluorescence stereomicroscope (Fig. 2.4b).

While they are undoubtedly powerful tools in GM research, the presence of visual/scoreable marker genes in a commercial crop variety is not essential and it is accepted that they should be avoided when producing transgenic plants for food or animal feed.

Design and Construction of Genes for Introduction into Plants

It is a common misconception among the public and the general media that a gene can be taken from any source, introduced into a crop plant unchanged and work properly. The reality is somewhat different. The coding sequence of a gene that is sourced from a bacterium, an animal or a different plant species will usually be translated properly to make a protein in a GM plant. However, the regulatory sequences that determine when and where in the organism a gene is active are less likely to be recognized the more distantly related the source organism is from the GM plant. A bacterial gene introduced unchanged into a GM plant will not be active at all, and vice versa. This means that the coding region of the gene to be introduced into a plant is usually spliced together with regulatory sequences that will work in the plant to make what is called a chimaeric gene.

Figure 2.5 shows a schematic diagram of a chimaeric gene comprising a promoter from a wheat seed protein gene (*Glu-D1x*) attached to the coding sequence from the bacterial gene, *UidA*, and the terminator from the *Agrobacterium* nopaline synthase gene (*Nos*). As I described in the previous section, the *UidA* coding sequence encodes an enzyme called β-glucuronidase that produces a blue pigment from a colorless substrate. The production of the blue product when the substrate is supplied shows that the enzyme is present and that the gene must be active. The *Nos* gene is one of those that is introduced into a plant during infection by wild-type *Agrobacterium tumefaciens* and has evolved to function in plant cells. The terminator sequence ensures that RNA from the introduced gene is processed properly. The other regulatory sequences of the chimaeric gene are present in the promoter from the wheat gene.

When wheat plants are genetically modified with this gene, the gene is active in the same tissue and at the same time as the gene from which the promoter came. That is specifically in the major storage compartment of the seed (the endosperm) in the mid-term of seed development. This can be shown by sectioning seeds from the GM wheat plants and incubating the sections in a medium containing the substrate for β-glucuronidase (Fig. 2.4a). A blue color develops in the endosperm tissue from mid-development seeds but not in sections taken from anywhere else in the plant.

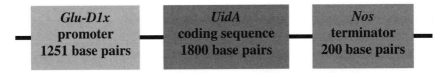

Fig. 2.5. Chimaeric genes.
Genes taken from one species and engineered into another are unlikely to work, particularly if the source and target species are not closely related. For this reason, chimaeric genes are constructed in the laboratory, comprising the coding region of the gene from the source species with control sequences (promoter and terminator) that will work in the target species. This figure shows a schematic diagram of such a chimaeric gene in which the coding region of the *UidA* gene from a bacterium is linked to a wheat gene promoter (*Glu-D1x*) that is active only in the seed endosperm and a terminator sequence from an *Agrobacterium tumefaciens* nopaline synthase (*Nos*) gene that functions in plants. Part of a section of a seed from a plant transformed with this gene is shown in Fig. 2.4a.

Promoter Types

The *Glu-D1x* promoter is known as a tissue-specific and developmentally-regulated promoter, being active in only one tissue type in the plant at a specific developmental stage. There are many different promoters of this type now available for plant biotechnology, and they are active in many different tissues and even cell types. For example, Fig. 2.6 shows β-glucuronidase activity in the roots, anthers and egg sacs of three different GM wheat plants, each containing the *UidA* gene under the control of a different promoter. It is an excellent demonstration of the control over the activity of a gene that is available to a biotechnologist. Such control is not possible with other methods in plant breeding. This tight control over promoter activity may be lost if the promoter is used in a different species, however, and this has to be tested before the promoter is used in commercial applications.

There are two other types of promoter available to the plant biotechnologist. The first type are known as constitutive, meaning that they are active everywhere in the plant all of the time (although many that are described as constitutive are much more active in some tissues than in others). In GM research, constitutive promoters are used most often to drive expression of selectable marker genes. However, the fact that they are expressed everywhere in the plant has also made them the promoter of choice for use in genes that impart herbicide tolerance or insect resistance in commercial GM crop varieties.

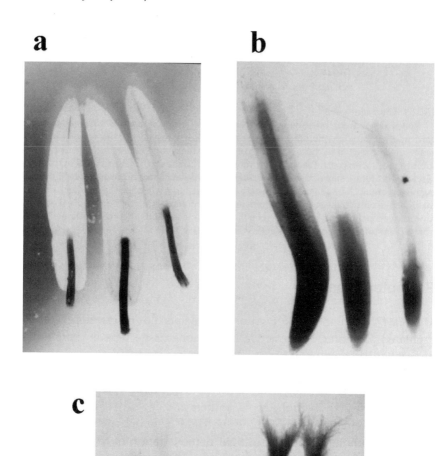

Fig. 2.6. Different promoters allow biotechnologists to specify where in a plant a gene will be active.

These pictures show the dark staining that indicates *UidA* gene activity in wheat plants that have been engineered with the gene under the control of different wheat promoters.

a. Activity in anther stalks.

b. Activity in different parts of the root.

c. Activity in different regions of the egg sacs.

Pictures kindly provided by Pilar Barcelo.

The most widely used constitutive promoter is derived from a plant virus, cauliflower mosaic virus. It is called the cauliflower mosaic virus 35S promoter, CaMV35S promoter for short. It has been used particularly successfully in the genetic modification of dicotyledonous plants. It also works in monocotyledonous plants, including cereal crop species, but has largely been replaced for cereal work with promoters that are derived from plant genes. These include a promoter for a maize gene, *Ubi*, encoding a protein called ubiquitin, and a rice gene, *Act1*, encoding a protein called actin. There has been some controversy over the use of the CaMV35S promoter (although not in the plant science community) because of its viral origins, even though the cauliflower mosaic virus only infects plants (specifically the cabbage family). The 35S promoter represents only a small part of the viral genome and could not in itself be infective.

The third type of promoter that a biotechnologist can use is described as inducible. Promoters of this type are not active until they are induced either by something such as attack by a pathogen, grazing or application of a chemical. There is some interest in using pathogen- or grazing-induced promoters in GM crops that are being modified to resist pathogen or insect attack. The resistance gene introduced into the plant would then only be active when it was needed. However, there are no reports of this being done successfully. Chemically-induced promoters are used in GM research so that a gene can be switched on and off in order to determine its function. There are no reports of their use in commercial GM crops and it is difficult to imagine how they could be used cost-effectively.

Examples of the Use of GM in Genetic Research

Another common misconception regarding plant genetic modification is that it is entirely concerned with producing new crop varieties for agriculture and is the preserve of big business. In fact, the techniques were developed in the public sector, largely in the US and Europe, and while there still remain only a small number of different types of genetic modification used in commercial crop plants (Chapter 3), the technique has been an extremely valuable tool in plant genetic research. The types of experiments to which it has been applied include analyses of gene promoter activity, functional characterization of regulatory elements within gene promoters, the determination of gene function, studies on metabolic pathways and analyses of protein structure and function.

The analysis of gene promoter activity is one way of finding out when and where in a plant a gene is active. The promoter of the gene is spliced to a reporter gene such as *UidA*, GFP or luciferase and the resulting chimaeric gene, known as a reporter gene construct, is introduced into a plant by genetic modification. The level and location of promoter activity can be visualized by testing for the presence of the protein produced by the reporter gene (see Figs. 2.4 and 2.6).

Further experiments can be undertaken to determine what parts of the promoter are important in its activity. This process often begins with an analysis of the DNA sequence of different promoters that have similar activity in order to find short regions of DNA (elements) that have the same sequence of base pairs. These elements are likely to play a role in controlling the activity of the promoter and are called regulatory elements. The function of potential regulatory elements can then be tested experimentally.

A good example of this was an analysis of the promoters of seed storage protein genes of wheat, barley, rye and maize. As their name suggests, these genes encode proteins that function as part of the storage reserve in the seed, providing nutrition for the developing seedling after germination. They have evolved into a large gene family, but are believed to have arisen from a single ancestral gene, and their activity is controlled in a co-ordinate manner. They are subject to tissue-specific and developmental regulation, being expressed exclusively in the major storage compartment of the seed, called the starchy endosperm, during mid and late seed development. They are also subject to nutritional regulation, responding sensitively to the availability of nitrogen and sulphur in the grain.

Since they have a common ancestry and show similar patterns of expression, it was to be expected that these genes would have regulatory sequences in common, and this turned out to be true for many of them. One group was found to contain a conserved element, twenty-nine base pairs long, positioned around 300 base pairs upstream of the coding sequence of the gene. This was one of the first plant regulatory elements to be characterized. It was identified by Brian Forde at Rothamsted in the UK and was first called the -300 element, subsequently the prolamin box. It has the sequence: TGACATGTAAAGTGAATAAGATGAGTCAT.

A regulatory role for the prolamin box was established experimentally by particle bombardment of cultured barley seed endosperms with promoter/*UidA* reporter gene constructs containing different regions of the promoter. These experiments also showed that the prolamin box could be subdivided into two separate elements, one (the E box) conferring tissue-specificity, the other (the N box) reducing activity of the gene at low nitrogen levels and increasing it when nitrogen levels were adequate. All of this control is imparted by the sequence of 29 base pairs.

Another valuable use of genetic modification in plant genetic research is in the analysis of gene function. These analyses are based on finding out what happens when a gene is switched off when it should be active (gene silencing) or what happens if more copies of a gene or a copy of a gene spliced to a more powerful promoter are introduced into a plant (gene over-expression).

There are three methods for silencing genes in plants using genetic modification. All three require the identification and cloning of the gene and the re-introduction of all or part of it into a plant. In the first method, a chimaeric gene is produced using part of the gene of interest in reverse orientation spliced downstream of a promoter sequence. The promoter may derive from the same gene, but usually it is a more powerful one. When this chimaeric gene is re-inserted into the plant, it produces RNA of the reverse and complimentary sequence of that produced by the endogenous gene. This so-called antisense RNA interferes with the accumulation of sense RNA from the target gene, preventing the sense RNA from acting as a template for protein synthesis. Use of a powerful promoter to drive expression of the antisense gene ensures that the antisense RNA is present in much greater quantities than the sense RNA.

The second method of gene silencing is called co-suppression and involves one or more additional copies of all or part of a gene in the correct orientation being introduced into a plant by genetic modification. In some cases, this leads to an increase in the production of the protein encoded by the gene. In others, however, gene silencing occurs and neither the native nor introduced gene is expressed.

Both co-suppression and antisense gene silencing have been used to produce genetically modified plants in which the trait is stably inherited. Both techniques have found commercial application in the extension of fruit shelf-life, most famously that of tomato (see Chapter 3). The co-suppression effect can be exerted either at the level of transcription, preventing the production of an RNA molecule from the DNA template, or post-transcriptionally, causing degradation of the RNA molecule before it can act as a template for protein synthesis. Post-transcriptional gene silencing (PTGS) turns out to be a defence mechanism against virus infection and some plant viruses have genes that suppress it. It involves the production of small, antisense RNAs, 25 nucleotides in length, that are specific for the transgene.

The production of these small, antisense RNA molecules can also be induced by the third method of gene silencing, RNA interference (RNAi). In this method, a plant is genetically modified to synthesize a double-stranded RNA molecule derived from the target gene. This has been done in other systems by splicing part of the gene between two opposing promoters and re-introducing it into an organism. It has been achieved in plants using gene constructs in which part of the gene is spliced sequentially in a head-to-tail formation downstream of a promoter. This causes the production of an RNA molecule that forms a hairpin loop (hpRNA) and is cleaved into 23 nucleotide-long RNA molecules. These induce enzymes called nucleases to destroy RNA produced by the target gene. It is still early days for RNAi but the degree of specific gene silencing obtained with this method appears to be much greater than that obtained using either co-suppression or antisense constructs. It also works in mammalian systems and is likely to have applications in medicine.

3 CURRENT AND FUTURE USES OF GM CROPS IN AGRICULTURE

Why Use GM in Plant Breeding?

The graph shown in Fig. 1.5 gives an indication of how successful plant breeding has been over the last century. So why do plant breeders need GM? The answer is that GM allows plant breeders to do some things that are not possible by other techniques. That does not mean to say that GM will replace older techniques in plant breeding, far from it, but GM is undoubtedly a powerful new tool for plant breeders. The advantages that GM has over other techniques are as follows:

(1) It allows genes to be introduced into a crop plant from any source. Biotechnologists can select a gene from anywhere in nature and with the modifications described in Chapter 2 make a version of it that will be active in a crop plant.
(2) It is relatively precise in that single genes can be transferred. In contrast, conventional plant breedinginvolves the mixing of tens of thousands of genes, many of unknown function, from different parent lines, while radiation and chemical mutagenesis introduce random genetic changes with unpredictable consequences.
(3) Genes can be designed to be active at different stages of a plant's development or in specific organs, tissues or cell types (see Chapter 2).

(4) Specific changes can be made to a gene to change the properties of the protein that it encodes.
(5) The nature and safety of the protein produced by a gene can be studied before the gene is used in a GM program (see Chapter 4).

GM also has some disadvantages. A successful GM program requires background knowledge of a gene, the protein that it encodes and the other genes and proteins that interact with it. This requires a significant investment of time and money compared, for example, with the generation of random mutations. Most significantly, though, GM varieties have to undergo much more detailed analysis and testing than new non-GM varieties, particularly in Europe. This probably explains why the few GM crop varieties available in Europe are essentially spin-offs from varieties developed elsewhere. The barriers to developing GM crop varieties specifically for the European market are too great. Even outside Europe, the GM crops that are currently available are based on a handful of strategies, although the number of different GM applications is growing.

Nevertheless, the GM crops that have been introduced onto the market have mostly been successful, some staggeringly so. By 2001 the worldwide area of land planted with GM crops exceeded 50 million hectares (124 million acres), approximately 6% of total world agriculture. More than half of the world's soybean crop was from genetically modified varieties. Countries where GM crops were grown in 2001 were US, Canada, Argentina, Uruguay, Mexico, Spain, Bulgaria, Ukraine, Romania, China, South Africa and Australia.

The National Center for Food and Agricultural Policy (NCFAP) of the US has reported on the effects of crop biotechnology in the US in 2001. The report states that the six crops currently in the marketplace (soybean, maize, cotton, papaya, squash and oilseed rape) produced an extra 1.8 million tonnes of food and fiber on the same acreage, improved farm income by US$1.5 billion and reduced pesticide use by 21 000 tonnes. The GM crops studied included insect-resistant maize and cotton, herbicide-tolerant soybeans, maize, cotton and oilseed rape, and virus-resistant squash and papaya.

Greatest yield increases were found with insect-resistant maize, while herbicide-tolerant soybeans gave the greatest cost savings through reduced herbicide use. GM crops also enabled farmers to adopt no-till, conservation systems of farming, reducing soil erosion. For example, 25 million acres of herbicide-tolerant soybeans were grown using this method in 2001.

These numbers are staggeringly large and their significance is clear. GM crops are bringing great benefits to farmers and the environment. Furthermore, the report predicted that the additional twenty-one crops in the pipeline would increase production by 4.5 million tonnes, improve income by US$1 billion and reduce pesticide use by 53 000 tonnes, if they were all adopted.

In this chapter, I will describe the different applications of GM in agriculture, and the potential applications of GM in the future.

Slow-Ripening Fruit

Much is made by opponents of GM of the notion that GM crop plants have been designed with farmers in mind instead of consumers. It would be reasonable to argue that there is nothing surprising or wrong in plant biotechnology companies trying to improve crops for farmers. Ironically, however, the first food on the market that was derived from GM plants were slow-ripening tomatoes with supposedly improved flavor, a consumer benefit.

Fruit ripening is a complex molecular and physiological process that brings about the softening of cell walls, sweetening and the production of compounds that impart color (lycopene), flavor and aroma. The process is induced by the production of a plant hormone, ethylene. The problem for growers and retailers is that ripening is followed sometimes quite rapidly by deterioration and decay and the product becomes worthless. Tomatoes and other fruits are, therefore, usually picked and transported when they are unripe. In some countries they are then sprayed with ethylene before sale to the consumer to induce ripening. However, fruit that is picked before it is ripe is widely believed to have less flavor than fruit picked ripe from the vine.

Biotechnologists saw an opportunity in delaying the ripening and softening process in fruit. By interfering with ethylene production or with the processes that respond to ethylene, fruit ripening could be slowed down. Fruit could then be left on the plant until it was ripe and full of flavor but would still be in good condition when it arrived at the supermarket shelf. Various strategies based on this principle are being pursued with many different fruits and the technology has the potential not only to improve the produce of Western farmers but also to enable farmers in tropical countries to sell fruit to customers in Europe and North America.

GM fruit that does not produce ethylene develops to the point where it would normally start to ripen and then stops. The farmer can then wait until all of the fruit has ripened and harvest it all at once. The fruit does not start to ripen until it has been sprayed with ethylene. Tomatoes of this type have been developed using antisense technology (Chapter 2) to decrease the production of the enzyme aminocyclopropane-1-carboxylic acid (ACC) synthase. This enzyme is responsible for one of the steps in ethylene synthesis and reducing levels of ACC decreases ethylene production dramatically. Tomatoes of this type were developed by a company called DNA Plant Technologies and are on the market in the US under the trade name "Endless Summer".

Another strategy that has been used with tomatoes is to add a gene that encodes an enzyme called ACC deaminase. This enzyme interferes with ethylene production by breaking down ACC. The gene was derived from a soil bacterium, *Pseudomonas chlororaphis*. Tomatoes of this type have been developed by Monsanto but are not yet on the market. A similar strategy used by Agritope, Inc., targets another of the precursors of ethylene, S-adenosyl methionine (SAM) and uses a gene encoding an enzyme called SAM hydrolase that breaks down SAM.

Another method of increasing the shelf-life of tomatoes targets the process by which fruit soften as they ripen. The GM fruit will change color and acquire aroma and flavor as normal but the softening process is slowed down. This means that the fruit are less prone to bruising and they deteriorate and decay much more slowly than normal. Fruit softening is brought about by enzymes that degrade complex carbohydrates in cell walls, including cellulases, which break down cellulose, and polygalacturonase (PG) and pectin methylesterase (PME), which are involved in the breakdown of pectin.

Two competing groups developed tomatoes with reduced fruit PG activity at approximately the same time. Calgene in the US used an antisense technique while Zeneca in collaboration with Don Grierson's group at the University of Nottingham used co-suppression (Chapter 2). The Calgene product was the Flavr Savr tomato, the first fresh fruit GM product on the market. It was introduced in 1996 but was not a commercial success and was withdrawn from the market after less than one year. Calgene was subsequently acquired by Monsanto, who have not so far pursued the technology further.

Zeneca chose to introduce the trait into a tomato used for processing and this proved to be much more successful. These tomatoes have a higher solid content than conventional varieties, reducing waste and processing costs in paste production and giving a paste of thicker consistency. This product went on the market in many countries and proved very popular in the UK from its introduction in 1996 until 1999 when most retailers withdrew it in response to anti-GM hostility.

Herbicide Tolerance

Weed control is an essential part of all types of agriculture and in the developed world the method of choice for most farmers to control weeds is to spray fields with chemical herbicides (weedkillers). This has been true since the 1950s, long before the advent of genetic modification. The alternatives are labor intensive and farmers could not abandon the use of herbicides with today's labor costs and keep food prices anything close to the level that they are now. Organic farmers do not use herbicides but organic farmers in the developed world are selling into niche markets, not providing the masses with affordable food. In the United Kingdom, for example, organic farmers simply avoid growing crops like sugar beet that are particularly sensitive to weed competition.

Most herbicides are selective in the types of plant that they kill and a farmer has to select a herbicide that is tolerated by the crop that he is growing but kills the problem weeds. Commercial (non-GM) sugar beet varieties grown in the UK, for example, are tolerant of around eighteen commercially available herbicides. By using a combination of different herbicides (typically eight) at different times in the season when different weeds become a problem, the farmer can protect his crop. Without weed control, sugar beet yield falls by about three-quarters and the crop is not worth harvesting.

This provides the farmer with a number of problems. The herbicide regime that he needs to use may be complicated. Some of the herbicides have to go into the ground before planting; once the crop seed is in the ground it is too late and weed problems cannot be responded to. Some of the herbicides involved are toxic to humans, are dangerous to handle and require protective clothing and equipment to be used. All herbicides require equipment and labor to apply them and some herbicides persist in the soil from one season to the next, making crop rotation difficult. Many of these problems have been overcome by the introduction of GM crops that tolerate broad-range herbicides. The first of these to be introduced were Roundup-Ready soybeans produced by Monsanto and marketed since 1996.

Roundup is Monsanto's trade name for glyphosate, a broad range herbicide that was introduced as a commercial product by Monsanto in 1974. It is now marketed under many different trade names in many agricultural and garden products. Glyphosate is non-selective and prior to its use in combination with GM plants it was used primarily to clear fields completely and remove weeds from pathways. It does not persist long in the soil because it is broken down by micro-organisms. How long it persists depends on soil type and typically ranges from a few days to several months. This means that many farmers can clear their fields and plant a crop a few weeks later. It would be extremely rare for glyphosate to persist at effective levels from one season to the next.

Glyphosate is taken up through the foliage of a plant, so it is effective after weeds have become established. Its target is an enzyme called 5-enolpyruvoylshikimate 3-phosphate synthase (EPSPS). This enzyme catalyzes the formation of 5-enolpyruvoylshikimate 3-phosphate (EPSP) from phosphoenolpyruvate (PEP) and shikimate 3-phosphate (S3P). This reaction is the penultimate step in the shikimate pathway, which results in the formation of chorismate, which in turn is required for the synthesis of many aromatic plant metabolites. These aromatic compounds include the amino acids phenylalanine, tyrosine and tryptophan, so amongst many other things plants treated with glyphosate are unable to make proteins. The herbicide is transported around the plant through the phloem and the whole of the plant dies.

The shikimate pathway is not present in animals, so animals have to acquire its products through their diet. This means that glyphosate has very little toxicity to insects, birds, fish or mammals, including man. For this reason farmers have always been very comfortable in using it.

Biotechnologists saw an opportunity to enable farmers to use glyphosate to control weeds simply and safely on growing crops by genetically modifying crop plants to tolerate the herbicide. They concentrated their efforts on EPSPS, initially trying to overcome the effect of glyphosate by engineering plants to overproduce the enzyme. Real success came with the discovery of a variant form of EPSPS (class II) in strains of the soil bacteria, *Agrobacterium tumefaciens* and *Achromobacter*. These enzymes worked efficiently but were not targeted by glyphosate. The *Agrobacterium tumefaciens* gene responsible was isolated and inserted into soybean by particle bombardment (Chapter 2).

The CaMV35S promoter (Chapter 2) was used to control the activity of the gene. The use of this promoter meant that the whole of the plant would be protected. The result was a variety of soybean that would tolerate glyphosate.

The first glyphosate-tolerant soybean line was called 40-3-2. Glyphosate treatment inhibits the plant's own EPSPS enzyme but does not affect the introduced enzyme. Line 40-3-2 was field tested in 1992 and 1993 and gave a similar yield after glyphosate treatment to untreated non-GM varieties. It has since been incorporated into many soybean breeding programs and over one hundred and fifty US seed companies now offer varieties carrying the trait.

Glyphosate can be applied to GM glyphosate-tolerant soybeans at any time during the season. This gives farmers flexibility in weed control and the ability to respond to a weed problem if it occurs unexpectedly. The regime that is used depends on the weed pressure in a particular area. In some cases a single application of glyphosate will control weeds throughout an entire growing season. However, the timing of the application is crucial if this is to work. Many farmers prefer to use a different herbicide before sowing in combination with a later application of glyphosate. Others use two sequential glyphosate applications.

The yield obtained with glyphosate-tolerant soybeans when they were undergoing trials was the same as for the non-GM controls. There were reports of disappointing yields in some areas when the first varieties were grown commercially, perhaps because the varieties that were available were not ideal for some locations. These reports have dried up as the trait has been bred into more varieties. There are also reports of yield increases in some areas.

The take-up of glyphosate-tolerant soybean varieties was incredibly rapid. After their introduction in 1996, their use rose rapidly to well over half of all the soybeans planted in the US in 2001, the total area of which was approximately thirty million hectares. Farmers cite simpler and safer weed control and reduced costs as the main reasons for using them. The exact economic impact is difficult to gauge. As glyphosate-tolerant varieties became more popular, sales of glyphosate rose while those of other herbicides, such as imazethapyr, acifluorfen, bentazon and sethoxydim, fell. The price of glyphosate was then reduced to maintain competitiveness. Overall, between 1995 and 1998 there was estimated to be a reduction of US$380 million in annual herbicide expenditure by US soybean growers (source: National Council for Food and Agricultural

Policy). However, farmers who used glyphosate-tolerant varieties had to pay a technology fee of US$6 per acre. This reduced the overall cost saving to US$220 million.

The technology fee did cause some resentment amongst US farmers, particularly when it was not imposed on their competitors in Argentina, who adopted glyphosate-tolerant varieties with even more enthusiasm than American farmers.

Another advantage for farmers using glyphosate-tolerant varieties is that crop rotation is made much easier. Some of the herbicides used on conventional soybean crops, such as chlorimuron, metribuzin, imazaquin and imazethapyr, remain active through to the next season and beyond. Some crops cannot be planted even three years after treatment. Maize, which is typically used in rotation with soybean, requires an eight to ten month gap before it can be planted. There are no such problems after glyphosate treatment because glyphosate is degraded so rapidly in the soil.

The use of glyphosate-tolerant soybean varieties has also allowed farmers to switch to a conservation tillage system, leaving the soil and weed cover undisturbed over winter. This reduces soil erosion and leaching of nitrate from the soil, where it is a fertilizer, into waterways where it is a pollutant.

Glyphosate tolerance has now been engineered into many crop species and commercial varieties of cotton, oilseed rape, maize, sugar beet and fodder beet are already on the market. It is undoubtedly the most successful GM trait to be used so far. Alternatives to the glyphosate-insensitive EPSPS enzyme are being explored. Several bacteria have been found to make an enzyme called glyphosate oxidoreductase that breaks glyphosate down and detoxifies it and genes encoding this enzyme have been engineered into several crop species.

Glyphosate-tolerant varieties are not the only GM herbicide-tolerant varieties available. They face competition from varieties that are tolerant of another broad-range herbicide, gluphosinate. These varieties were developed by Plant Genome Systems, which was subsequently acquired by Aventis, who themselves were acquired by Bayer. Gluphosinate (or glufosinate) is marketed under the trade name Liberty. The gene used to render plants resistant to it comes from the bacterium *Streptomyces hygroscopicus* and encodes an enzyme called phosphinothricine acetyl transferase (PAT). This enzyme detoxifies gluphosinate. Crop varieties carrying this trait carry the trade name Liberty Link and include varieties of oilseed rape, maize, soybeans, sugar beet, fodder beet, cotton and rice. The oilseed rape variety has been particularly successful in Canada.

Insect Resistance

Another universal problem for farmers all over the world is loss of their crop to insect grazing. Extreme examples of this are locust swarms in Africa which can destroy the crops of entire areas and the Colorado beetle, which in bad years destroys 70% of the potato harvest in Eastern Europe and Russia. Farmers in developed countries usually use chemical controls and to a lesser extent biological controls (such as natural predatory species) to prevent these sorts of losses. However, these measures are expensive, the pesticides used are often toxic and difficult to handle safely, while biological controls are never 100% effective.

Organic and salad farmers have been using a pesticide called Bt for several decades. It actually consists of a soil bacterium, *Bacillus thuringiensis*, which produces a protein that is toxic to some insects. Bt pesticides are applied as powders, granules, or aqueous and oil-based liquids. Organic farmers use Bt pesticides because they are biological products that degrade rapidly. Salad farmers use them because they can be applied immediately before harvest. They have no toxicity to mammals, birds or fish and have an extremely good safety record. Disadvantages are that they are specific for insect types and do not remain effective for long after application.

The protein that *Bacillus thuringiensis* produces that is toxic to insects is called the Cry protein. Different strains of the bacterium produce different versions of the protein, and these versions are classified into groups CryI–CryIV. Each group is subdivided further into subgroups A, B, C, etc. The different proteins are effective against different types of insects. CryI proteins, for example, are effective against the larvae of butterflies and moths, while CryIII proteins are effective against beetles.

Biotechnologists took the view that engineering crops to produce the Cry protein would be more efficient than applying the protein externally. This would overcome the problem of rapid loss of activity after application and would also have the advantage that only insects eating the crop would be affected. The *Cry1A* gene from several strains of *Bacillus thuringiensis* has now been introduced into crops, including cotton, sugar beet and maize. The modified varieties are generally referred to as Bt varieties, although different companies market them under a variety of trade names. The effect of the use of Bt cotton has perhaps been the most striking. Conventional cotton is very susceptible to insect damage and one quarter of US insecticide production is used on this one crop. In areas of severe pressure from the three major pests affected by the CryIA protein (tobacco budworm, cotton bollworm and pink bollworm), Bt cotton on average requires 15–20% of the insecticide used on conventional cotton and take-up of Bt varieties in these areas has been very high. In Alabama, for example, Bt varieties accounted for 77% of the cotton planted in the first year that they became available. In areas where Cry1A-controlled pests are less abundant, adoption rates of the new cotton varieties have been low.

The effects of adoption of Bt maize have varied from area to area. The principle maize pest affected by the Cry1A protein is the corn borer and where the corn borer is not particularly abundant, US farmers do not spray against it, preferring to tolerate any losses incurred if an infestation occurs. In these areas use of Bt maize has lead to an increase in yield typically of 10–15%. In areas where the corn borer is abundant, farmers have had to spray heavily and pre-emptively with insecticide to control it. In these areas, use of Bt maize has not increased yields but has lead to a decrease in the number of sprays applied per season from an average of 7–8 to an average of 1–2.

An unexpected benefit of using Bt maize varieties is that the Bt grain contains lower amounts of fungal toxins (mycotoxins). These compounds, which include potent carcinogens such as aflatoxin and fumicosin, are a particular problem in tropical countries because of the warm and humid conditions that maize grain is stored in. They are associated with a high incidence of throat cancer. In temperate countries they are considered to pose most risk to grain-fed animals and the people most aware of them are horse-owners. Nevertheless, there is undoubtedly a low-level presence of these chemicals in the human food chain. The reduced insect damage done to Bt crops means that they are less susceptible to fungal infection.

The other *Cry* gene that has been used in plant biotechnology is the *CryIIIA* gene of *Bacillus thuringienis* var. *tenebrionsis*. The CryIIIA protein is effective against beetles and potato varieties containing the *CryIIIA* gene are resistant to infestation by the Colorado beetle, a beetle that can devastate potato crops. One such variety, NewLeaf, produced by Monsanto, has been on the market in the US for several years but has recently been withdrawn due to poor sales. US potato plantations are attacked by a number of pests that are not controlled by Bt and farmers have turned to new, broad-range insecticides instead of the GM option. This decision was also undoubtedly influenced by the MacDonalds burger chain, which refused to accept the new variety (MacDonalds is extremely reactionary in its attitude to new potato varieties, GM or otherwise). Nevertheless, Bt potato may have a role to play in Eastern Europe and Russia where the Colorado beetle is a huge problem.

Virus Resistance

Plant viruses could be regarded as being as dangerous to humans as viruses causing the worst human diseases. Viruses such as Cassava Mosaic Virus and the Feathery Mottle Virus of sweet potato, for example, are estimated to be responsible for the deaths of millions of people every year through the destruction of vital food crops.

Farmers attempt to prevent viral plant diseases by controlling the insect pests that carry the disease. Farmers in developed countries also have access to virucidal chemical applications but, again, these are expensive and, once established, viral diseases are difficult to bring under control.

The first methods used by biotechnologists to engineer plants to be resistant to viruses arose from investigations into cross protection, in which infection by a mild strain of a virus induces resistance to subsequent infection by a more virulent strain. Cross protection has been known about for some time but exactly how it works in plants is not understood (plants do not have an immune system like that of animals). It appears to involve the coat protein of the virus because encapsulating a virus in the coat protein of a different virus eradicates it. Engineering a plant to make a viral coat protein was found to mimic cross protection, the plants showing delayed or no symptoms after infection.

Another method to engineer virus resistance is to use antisense or co-suppression (Chapter 2) to block the activity of viral genes when the virus infects a plant. Monsanto, for example, used a replicase gene from potato leaf role virus (PLRV) to induce resistance to PLRV in potato. A potato variety containing this trait and the Bt insect-resistance trait was marketed under the trade name NewLeaf Plus. Other applications include inducing resistance to Spotted Wilt Virus in tomato.

Another example of the commercialization of virus-resistant GM varieties is papaya grown in the Puna district of Hawaii. After an epidemic of papaya ringspot virus (PRSV) almost destroyed the industry, growers switched in 1998 to a virus-resistant GM variety. The GM variety contains a gene that encodes a PRSV coat protein. There is no other known solution to an epidemic of PRSV and the GM variety probably saved the papaya industry in Hawaii.

Perhaps the most exciting examples of the use of this technology are in developing countries where losses to viral diseases are the greatest and have the most severe consequences. Kenya is currently testing a GM sweet potato variety that is resistant to Feathery Mottle Virus. This virus is estimated to reduce Kenya's sweet potato production by half and the yield of the GM variety is reported to be 80% higher than non-GM varieties in the trials.

Modified Oil Content

The principle components of plant oils are fatty acids and the various properties of oils from different plants are determined by their differing fatty acid contents. Many hundreds of different fatty acids have been identified in plants. Each fatty acid comprises a carboxylic acid group at the end of a hydrocarbon chain, the length of which in plant fatty acids ranges from eight to twenty-four carbons. Fatty acids are differentiated further in the number and position of double (or unsaturated) bonds between the carbons in the chain.

Different fatty acids can be purified from the oils of plants and have diverse uses. Lauric acid (twelve carbons), for example, is used in cosmetics and detergents. Palmitic acid (sixteen carbons), stearic acid (eighteen carbons) and oleic acid (eighteen carbons, one unsaturated bond) are used in foods, while gamma-linolenic acid (eighteen carbons, three unsaturated bonds) is used in health products. Erucic acid (twenty-two carbons, one unsaturated bond) is poisonous but is used in the manufacture of plastics and lubricating oils.

Some plant seeds accumulate oils as a storage reserve in their seeds and it is the seeds of oil crops that are harvested and from which oil is extracted. A major oilseed crop in North America and Europe is oilseed rape (*Brassica napus*). Oilseed rape was used as a forage crop in the nineteenth century. By the early twentieth century it had been found that plants could be selected and crossed to produce an oil crop. However, the oil contained high levels of erucic acid and the meal left over after oil extraction contained high levels of compounds called glucosinolates. Both erucic acid and glucosinolates are poisonous. The first time that oilseed rape was grown widely in the United Kingdom was during the Second World War, when it was used to produce industrial oil. High erucic acid varieties are still grown today for this purpose but are not permitted to be placed in the food chain. Plant breeding of different varieties over the next thirty years reduced the levels of erucic acid and glucosinolates to the point where these varieties were considered acceptable for human consumption.

The first low erucic acid, low glucosinolate varieties were grown in Canada in 1968. Nevertheless, oilseed rape did not get its seal of approval (Generally Recognized as Safe) from the Food and Drug Administration of the US until 1985. Canadian producers then came up with the name Canola (derived from Canadian oil) for edible oilseed rape oil. This name was adopted all over North America as the name not only for the edible oil but also for the crop itself.

Oilseed rape oil is one of the cheapest edible oils on the market. In order to make a more valuable oil from oilseed rape, the company Calgene, subsequently taken over by Monsanto, genetically modified an oilseed rape variety to produce high levels of lauric acid in its oil. This variety was introduced onto the market in 1995. It contains a gene from the Californian Bay plant that encodes an enzyme that causes premature chain-termination of growing fatty acid chains. The result is an accumulation of the twelve carbon chain lauric acid to approximately 40% of the total oil content, compared with 0.1% in unmodified oilseed rape. Lauric acid is a detergent traditionally derived from coconut or palm oil. The advantage for farmers growing the new variety is that they get a premium price for their product.

The other major crop that has been genetically modified to increase the value of its oil is soybean. The genetically modified variety was produced by PBI, a subsidiary of DuPont. It accumulates oleic acid, an eighteen carbon chain fatty acid with a single unsaturated bond (a monounsaturate) to approximately 80% of its total oil content, compared with approximately 20% in non-GM varieties.

In conventional soybean much of the oleic acid is converted to linoleic acid, an eighteen chain fatty acid with two double bonds (a polyunsaturate), by an enzyme called a delta-12 desaturase. Some of the linoleic acid is further desaturated to linolenic acid, a polyunsaturate with three double bonds. In the GM variety, the activity of the gene producing this enzyme is reduced by co-suppression (see Chapter 2) so that oleic acid levels are increased while linoleic and linolenic acid levels are decreased.

Oleic acid is very stable during frying and cooking and is less prone to oxidation than polyunsaturated fats, making it less likely to form compounds that affect flavor.   The traditional method of preventing polyunsaturated fat oxidation involves hydrogenation and this runs the risk of creating *trans*-fatty acids. *Trans*-fatty acids contain double bonds in a different orientation to the *cis*-fatty acids present in plant oils. They behave like saturated fat in raising blood cholesterol and contributing to blockage of arteries. The oil produced by high oleic acid GM soybean requires less hydrogenation and there is less risk of *trans*-fatty acid formation.

As with the high lauric acid oilseed rape, the advantage to farmers of growing high oleic acid soybean is that they get a premium price for it.

Prospects — Crops on the Way

The GM crops being grown commercially today involve the application of a tiny fraction of the possibilities that genetic modification of plants offers. Here are a few examples of GM crops that might appear on the market in the next few years.

Production of Industrial Oils and Pharmaceutical Fatty Acids

The examples that I have just described are undoubtedly only the first of many GM crops modified to change their oil content. Many oils have uses in the plastics and industrial oil industries and there are examples of crops that have been modified to make oils of this type. At present, these oils still have to compete with cheap petroleum derivatives but are likely to become increasingly valuable when supplies of petroleum eventually start to run out. It should be noted, however, that at present yields these crops could not come close to satisfying the global demand for oil and other alternatives to petroleum as a source of fuel and a raw material will have to be found as well.

An even more exciting prospect is the production of oils with nutritional or pharmaceutical properties. Animals, including humans, have only a limited capacity to synthesize fatty acids and a number of fatty acids produced by plants have been identified as essential dietary components (essential fatty acids, or EFAs). Furthermore, a number of fatty acids produced by plants have pharmaceutical properties, including gamma-linolenic acid (GLA), which is found in borage and evening primrose, and arachidonic acid (AA) which is only found in a few mosses and fungi. GLA is used in the treatment of skin conditions such as atopic eczema and also has anti-viral and anti-cancer properties. AA is a constituent of breast milk and is important for brain and eye development in infants.

The plant species that produce these fatty acids at high levels make very poor crops. A target for biotechnologists, therefore, is to take the genes that encode the enzymes responsible for making these fatty acids and engineer them into crop plants. This has already been shown to work in Johnathan Napier's laboratory at Long Ashton Research Station in the United Kingdom. A gene from borage encoding an enzyme called a delta-5 desaturase was introduced into transgenic plants and these plants were shown to accumulate GLA. The host plant has to make the precursor for GLA so that the introduced enzyme has a substrate to work on. However, sunflower and flax, both established crop plants, contain this precursor.

GM plants modified to produce these oils could be used as a source for the pharmaceutical product. Alternatively, some could be used as supplements in so-called neutraceutical foods.

Vitamin Content

Vitamins are compounds that are required for the normal functioning of our bodies but which we cannot make ourselves. They are, therefore, an essential component of our diet. They are involved in physiological processes such as the regulation of metabolism, the production of energy from fats and carbohydrates, the formation of bones and tissues and many others.

While nutritionists still argue over the optimum levels of different vitamins that people should consume, it is agreed that consumers in the developed world have access to foods (fresh fruit, vegetables, bread, meat, dairy products) containing all of the vitamins that they require. They also have access to an array of vitamin supplements. Most people eat plenty of meat and dairy products, but a significant number avoid them altogether. Few people eat the recommended amounts of fresh fruit and vegetables. The health of some people might, therefore, be improved if the levels of vitamins in foods were increased. In other words put vitamins in foods that consumers like instead of persuading consumers to change their diets. Furthermore, for foods that consumers associate with a healthy lifestyle, such as breakfast cereals, there is market advantage to be gained in improving the nutritional value of the product and advertising the fact to consumers.

Obviously, manufacturers have the option of adding vitamins to their products but in some cases this is expensive. This might present opportunities for biotechnologists to engineer crop plants to contain more vitamins. Folic acid, deficiency of which may cause gastrointestinal disorders, anemia and birth defects, is one possible target. Others are the fat-soluble vitamins E and K, deficiencies in which are associated with arterial disease and, in the case of vitamin K, post-menopausal osteoporosis.

The more obvious need for increasing the vitamin content of people's diets is in the developing world, where acute vitamin deficiency is the cause of poor development, disease and death. Vitamin A deficiency, for example, is common in children in developing countries who rely on rice as a staple food. It causes symptoms ranging from night blindness to those of xerophthalmia and keratomalacia, leading to total blindness. It is the leading cause of blindness in children in the developing world. For example, it is estimated that a quarter of a million children go blind each year because of vitamin A deficiency in Southeast Asia alone. Vitamin A deficiency also exacerbates diarrhea, respiratory diseases and measles; improving vitamin A status in children reduces death rates by 30–50%.

The World Health Organization aimed to eradicate vitamin A deficiency by the year 2000, but failed abjectly due to an inability to reach those in need. A biotechnologist, Ingo Potrykus, from the Swiss Federal Institute of Technology in Zurich, saw the possibility of using an alternative approach through the genetic modification of rice. Rice grain does contain vitamin A but only in the husk. The husk is discarded because it rapidly goes rancid during storage, especially in tropical countries. Potrykus aimed to produce a rice variety that would make pro-vitamin A (beta-carotene, a precursor that humans can process into vitamin A) in its seed endosperm, the largest part of the seed that is eaten. A conventional breeding approach would not work because no conventional rice variety makes any pro-vitamin A at all in its endosperm.

Rice endosperm synthesizes a compound called geranylgeranyl diphosphate, which is an early intermediate in the pathway for beta-carotene production. Potrykus' team successfully engineered the rest of the pathway into rice, using phytoene synthase (*psy*) and lycopene β-cyclase genes from daffodil (*Narcissus pseudonarcissus*), and a phytoene desaturase (*crtI*) gene from the bacterium *Erwinia uredovora*.

The GM rice producing pro-vitamin A was crossed with another line containing high levels of available iron. Rice normally contains a molecule called phytate that ties up 95% of the iron, preventing its absorption in the gut. The GM rice contains a gene encoding an enzyme called phytase that breaks phytate down. The high pro-vitamin A/high available iron hybrid was called Golden Rice.

The aim of the Potrykus team is to enable people to obtain sufficient pro-vitamin A to avoid the symptoms of vitamin A deficiency in a daily ration of 300 grams of rice. The new "Golden Rice" has yet to be crossed with commercial rice varieties and it remains to be seen what levels of pro-vitamin A will accumulate in the products of these crosses. Nevertheless, the potential of the work is extremely exciting. The Rice Research Institute in Manila, The Philippines, is already crossbreeding Golden Rice with its local rice varieties and breeders in India are set to begin doing the same.

Fungal Resistance

Fungal diseases of plants cause severe losses in crop production. An extreme example is the Irish potato famine of the nineteenth century, which was caused by a pandemic of the fungus *Phytophthora infestans*, which causes late blight disease. Resistance to infection on the part of the plant is imparted by so-called resistance or R genes. R genes encode proteins that act as receptors for pathogen molecules and induce a hyper-sensitive response (HR) in which there is rapid cell death around the entry point of the fungus, preventing development of the disease.

The pathogen proteins that are detected by R genes are encoded by avirulence genes (*Avr* genes). An *Avr* gene aids infection of plants that do not carry an R gene that recognizes it but must be discarded by the pathogen to overcome resistance imparted by an R gene. This complicated relationship between the genetics of plants and their fungal pathogens is the product of millions of years of co-evolution.

The success of R genes is dependent on there not being too many individuals carrying any one particular R gene. There is then some selective advantage in a pathogen retaining an *Avr* gene, in that it is better able to infect those plants without the R gene. In a field of crop plants that are essentially all the same this selective advantage does not exist. As a result, the efforts of plant breeders to breed resistance into crop varieties by introducing particular R genes have rapidly been overcome by the emergence of new strains of the pathogen that are not affected. Nevertheless, biotechnologists believe that if they can learn more about how the R gene system works they will be able to engineer fungal resistance into crop plants, perhaps by stacking multiple R genes in one crop variety.

An alternative approach to the challenge of engineering fungal resistance into crop plants is to modify them with genes that express fungicidal proteins. Examples are genes encoding the enzymes chitinase and β-glucanase, both of which attack the cell walls of fungal hyphae as they enter the plant. Transgenic plants containing genes for these enzymes have been reported to have increased resistance to pathogenic fungi but there is no commercial use of this technology that I am aware of.

Salt Tolerance

Agricultural production is severely affected by the presence of high concentrations of salt in soil and the effects of salt building up in soils have been a problem for farmers throughout history, from ancient Egypt to modern-day US. Today, millions of acres of otherwise fertile land in developed and developing countries are rendered useless by salt build-up. The most common cause is irrigation, where the small amounts of salt in river water are concentrated by evaporation and build up in the soil. Engineering salt tolerance into crops through genetic modification is, therefore, an important target for plant biotechnologists.

Plant cells contain a compartment called the vacuole in which excess salt can be dumped where it will not affect the workings of the rest of the cell. This requires a special protein called a vacuolar Na^+/H^+ antiport that pumps the salt from the main part of the cell into the vacuole. Eduardo Blumwald and co-workers at the University of California have produced

GM tomato plants that make more of these molecular pumps. The GM tomato plants can tolerate salt concentrations several times higher than non-GM plants and should survive comfortably in the salt concentrations of soils that are currently considered unusable. Salt accumulates in the leaves but the fruit remains edible and not salty at all. This means that removal and disposal of the leaf material after harvest actually cleans up the soil and a few harvests of the GM plants should return the soil to salt concentrations suitable for growth of other crops.

Edible Vaccines

It is estimated that there are twelve billion injections administered worldwide every year. Thirty percent of these are not performed under sterile conditions. Furthermore, millions of people die every year from diseases that are preventable through vaccination. They do not have access to vaccines for one or more of several reasons. They may live in remote places that lack the infrastructure (transport, refrigeration, availability of needles and syringes) to deliver and store the vaccine. People will often walk for many days to a clinic to have their children and themselves vaccinated but find it impossible to repeat the journey for a subsequent booster vaccination, so they are not protected adequately. There may not be sufficient numbers of medical professionals to administer a vaccination program or the vaccine may just be too expensive.

A vaccine for which all of these factors apply is that for hepatitis B. Hepatitis B causes acute and chronic disease of the liver and is associated with liver failure and liver cancer. It is a big killer throughout the developing world. The first hepatitis B vaccine was a protein (the surface antigen) extracted from the blood of people infected with the disease and was produced in the 1970s. As HIV spread in the 1980s, this practice was viewed as too dangerous and the gene that encoded the protein was engineered into GM yeast to produce the vaccine. The vaccine has been made this way ever since and is extremely effective. Unfortunately it is far too expensive for developing countries to afford and even in the United Kingdom the vaccine is only offered to those people who are considered to be at special risk of contracting the disease.

A group of researchers lead by Hugh Mason at Cornell University genetically modified potato plants to make the Hepatitis B surface antigen in their tubers. The vaccine is administered orally (patients eat a piece of the GM potato) and has been shown to work in animal and human trials. The original intention with this product was that people would be able to grow it themselves. However, this would make the control of dosage impossible and is not now considered to be a realistic option. Nevertheless, the advantages are clear: the potato can be stored at room temperature for a long time, and even moderate-scale production by agricultural standards could supply enough vaccine to meet global demand. Similar experiments have been performed with a protein from pathogenic *Escherischia coli*, a bacterium that causes severe, sometimes fatal food poisoning. Human trials of this vaccine were completed in 1997.

The potential of both of these products seems huge. However, perhaps because of their novelty, the fact that they come from GM plants (instead of GM yeast, ironically) or a lack of will to develop products that would bring most benefit to people of developing countries, they have not yet been taken up by the pharmaceutical industry.

Antibody Production in Plants

Vaccines induce the immune system to produce antibodies. An alternative strategy being pursued by biotechnologists is to engineer plants to make antibodies themselves. Plants are complex, highly evolved organisms, equivalent to higher animals such as ourselves in that respect. They can synthesize and process proteins such as antibodies correctly in a way that yeast and other micro-organisms cannot. An application of this technology is in the prevention of tooth decay. Tooth decay is caused by a bacterium called *Streptomyces mutans*. It has proved possible to produce effective monoclonal antibodies against this bacterium in plants with the aim of providing protection against this bacterium in a mouth-wash or toothpaste. However, once again this awaits development by the pharmaceutical industry.

Removal of Allergens

Food allergy is an increasingly serious medical problem. The Food Standards Agency of the United Kingdom reported in 2000 that around 1.4–1.8% of the UK population, and up to 8% of children in the UK suffer from some type of food allergy. It appears that there have been rapid increases in the number of incriminated foods and the frequency of severe reactions. At present there is no specific treatment for food allergy, apart from dietary avoidance, although patients at risk of anaphylactic shock carry adrenaline in case of accidental exposure. Avoidance of allergens in processed foods can be difficult because their presence may not be obvious and products may not be labeled properly. The risk of accidental exposure, for example in a restaurant meal, is a constant anxiety that affects quality of life significantly.

Ironically, the possibility that some genes used in plant biotechnology could encode allergenic proteins has concerned consumers and regulatory authorities. This is considered in Chapter 5. The other side of the coin is that genetic modification could be used to remove allergens from the food chain.

Well known plant allergens include the 2S albumins, a group of storage proteins found in the seeds of legumes (peas and beans), crucifers (cabbages, turnips, swedes, mustards, radish) and many nuts. As well as acting as food stores for the seed, some inhibit digestive enzymes, perhaps protecting the seeds from being eaten, and some have antifungal properties.

Another group of proteins that includes several allergenic members are lipid transfer proteins, a family of small proteins present in seeds and other parts of a diverse range of plants, including fruits, oilseeds and cereals. They cause sunflower allergenicity in Southern European populations, resulting in severe symptoms, including anaphylaxis, albeit in a small number of sufferers. The function of lipid transfer proteins is not known but they have been associated with plant defense.

Another family of proteins involved in plant defense are the PR (pathogen-related) proteins. These are made by plants in response to microbial infection. They include a number of allergenic proteins and are responsible for the allergenicity of chestnut, avocado, birch pollen, apples, cherries, celery and carrots.

Baker's asthma, a respiratory allergy caused by inhalation of wheat flour, is common in workers in the flour milling and baking industries. The allergenic proteins responsible appear to be small inhibitors of two digestive enzymes, alpha-amylase and trypsin. They are present in rye, barley, maize and rice as well as wheat.

The obvious target for biotechnologists is to remove these proteins from crop plants using antisense, co-suppression or RNA inhibition techniques (Chapter 2). These methods can lead to almost total suppression of gene activity and, while it might not be sufficient to make a food completely safe for someone who is allergic to it, it could reduce the frequency and severity of reactions caused by accidental exposure. However, clearly this can only be done if the protein does not have an essential function in the plant.

There are currently only a few examples of experiments in which this has been tried. In one, the amounts of alpha-amylase inhibitors in rice were reduced substantially using the antisense method. There are at least ten different genes encoding these proteins in rice and to affect them all using conventional plant breeding or mutagenesis would be impossible. In another experiment, antisense technology was used to reduce the levels of the Lol p5 allergen in the pollen of ryegrass. The pollen of the GM ryegrass was shown to have reduced allergenicity.

The list above is not meant to be an exhaustive one. Genetic modification could be used to produce pharmaceuticals, fragrances and pigments, increase nutritional value in many different ways, improve the processing properties of crop plants (for example the breadmaking quality of wheat), address agronomic problems such as premature sprouting of grain and tubers and pod shatter in oilseed rape, and reduce chemical inputs (and therefore cost) in agriculture even further. These advances will only be made, however, if the legislation covering the use of the technology does not make it too difficult or expensive to produce and market GM crops and consumers accept GM crop products.

4 LEGISLATION COVERING GM CROPS AND FOODS

Safety of GM Plants Grown in Containment

The question of GM crop and food safety was first considered in the United Kingdom in the early 1980s when the first GM plants were being produced. Responsibility for ensuring that the technology was developed safely was given to the Health and Safety Executive (HSE), a government agency responsible for the regulation of almost all of the risks to health and safety associated with the workplace in the United Kingdom. The HSE had already set up a committee, the Advisory Committee on Genetic Modification, to control the use of GM micro-organisms. The responsibility of this committee was extended to cover the production and use of all GM organisms (GMOs) in containment.

No organization in the United Kingdom can produce or hold GMOs without the permission of the HSE. The HSE ensures that any organization that proposes working on GMOs has the facilities required and has staff who are trained and experienced in the handling and disposal of the organisms and contaminated waste. Different levels of containment are required for different GMOs, depending on the risk that they represent. However, even for GMOs that are considered to be of no risk to human health or the environment, the laboratory in which they are held and used must satisfy basic standards. For example, the laboratory must be easy to clean, bench-tops and floor must be sealed and if the laboratory is mechanically ventilated the air flow must be inwards. Hand-washing facilities must be provided by the exit and basic protective clothing, such as a lab-coat and disposable gloves, must be worn and removed before leaving the laboratory. Access to the laboratory must be restricted. Any procedures that produce aerosols must be done in a safety cabinet, effective disinfectants must be available next to every sink, spillages must be dealt with immediately, bench tops must be cleaned after use and a good general standard of cleanliness maintained. All contaminated glassware must be stored safely and sterilized. All contaminated waste must be stored safely in a designated bin and autoclaved (sterilized by heat and pressure) before disposal.

At the heart of the safe handling of GMOs is expert risk assessment. All projects involving GMOs have to be risk assessed by the project leader and the risk assessment must be considered by an internal Genetic Modification Safety Committee. The project leader is required to give information on the experience level of staff who will work on the project and the training that they have received. For a project concerning GM plants, information has to be given on the host plant and the gene(s) being inserted. To assess any risk to human health the project leader has to consider any possible induction of or increase in toxicity and/or allergenicity compared with the parent plant and the risk of accumulation of toxicity through food chains.

The project leader also has to assess any risk posed by the proposed GM plants to the environment, particularly the potential of the plants to be more "weedy" than the parent plant. This assessment will include factors such as colonization ability, seed dispersal mechanisms, resistance to control measures such as herbicides, increased toxicity to insects and other grazers and any other possible change in the plants' interaction with their environment. The potential for and consequences of the sexual transfer of nucleic acids between the GM plants and other plants of the same species or a compatible species has to be considered, particularly if the plants have the ability to transfer novel genetic material to UK plant species. The risk and consequences of horizontal gene transfer to unrelated species, for example by a virus, bacterium or other vector, also has to be taken into account. Finally, an assessment has to be made of the potential of the GM plants to cause harm to animals or beneficial micro-organisms.

This risk assessment has to be completed before the experiment begins and despite the fact that the plants will be kept in containment in specially-designed greenhouses or controlled environment rooms. These have filtered negative air pressure ventilation, sealed drains and a chlorination treatment system for drainage water to ensure that no viable plant material escapes into the environment.

The proper risk assessment of experiments involving GMOs and their safe handling and containment are legal requirements and failure to observe the regulations could result in prosecution. It would certainly result in severe cases in the loss of an institution's licence to work with GMOs. All institutions that work with GMOs are inspected regularly by the HSE.

Safety of Field Releases of GM Plants

In the late 1980s it became clear that the growing of GM plants in the field in the United Kingdom was going to be widespread and the government decided that it would have to be regulated. Regulations covering the field release of GM plants and animals were written into the Environmental Protection Act of 1990 and a statutory advisory committee was set up to provide advice to government regarding the release and marketing of genetically modified organisms. This committee was called the Advisory Committee for Releases into the Environment (ACRE).

The regulations were updated through the GMO Deliberate Release Regulations 2002 which implement the European Union's directive on the deliberate release of GMOs, Directive 2001/18/EC.

ACRE advises the Secretary of State for Environment, Food and Rural Affairs (DEFRA), Scottish Ministers, the Welsh Assembly Secretaries and the Department of the Environment in Northern Ireland. It also advises the Health and Safety Executive on human health aspects of releasing GMOs.

ACRE is made up predominantly of academics. Current members have expertise in agronomy, biodiversity, conservation, ecology, entomology, genetics, microbiology, molecular biology, plant breeding, plant physiology, rural affairs, virology and weed ecology. The committee also includes two farmers.

Anyone who wants to undertake a field release of a GM plant in the UK has to obtain permission from DEFRA. A detailed risk assessment of the plants must be undertaken and considered by ACRE, who then advise DEFRA on whether or not to allow the release to go ahead. More information will be available for this risk assessment than the one described for a proposal to make a set of GM plants and grow them in containment because the plants will already have been made and studied. A typical risk assessment will include:

- The full name of the plant.
- The breeding line used.
- Details of the sexual reproduction of the plant.
- Generation time.
- The sexual compatibility of the plant with other cultivated or wild plant species.
- Information concerning the survivability and dissemination of the plant.
- The geographical distribution of the species.
- Potential toxic effects on humans, animals and other organisms.
- A description of the methods used for the genetic modification.
- The nature and source of the vector used to modify the plant.
- Details of the novel genes introduced into the plant, including size, intended function and the organisms from which they originated.
- A description of the trait or traits and characteristics of the genetically modified plant which have been introduced or modified.
- The size and function of any regions of the host plant genome that have been deleted as a result of the modification.
- The location of the inserted novel DNA in the plant cells (usually integrated into a chromosome).
- How many copies of the novel gene or genes is/are present.
- Information on when and where in the plant the novel gene or genes is/are active and the methods used for finding out.
- Any differences between the genetically modified plant and its parent in respect of methods and rates of reproduction, dissemination and survivability.
- The genetic stability of the novel gene or genes.
- The potential for transfer of genetic material from the genetically modified plants to other organisms.
- Information on any toxic or harmful effects on human health and the environment arising from the genetic modification.
- The mechanism of interaction between the genetically modified plants and target organisms (for example if the plants have been engineered to be resistant to insects).
- Any potentially significant interactions with non-target organisms.
- A description of detection and identification techniques for the genetically modified plants.

- Information about previous releases of the genetically modified plants.
- The location and size of the release site or sites.
- A description of the release site ecosystem, including climate, flora and fauna.
- Details of any sexually compatible wild relatives or cultivated plant species present at the release sites.
- The proximity of the release sites to officially recognized protected areas that may be affected.
- The purpose of the release.
- The foreseen dates and duration of the release.
- The method by which the genetically modified plants will be released.
- The method for preparing and managing the release site, prior to, during and after the release, including cultivation practices and harvesting methods.
- The approximate number of genetically modified plants (or plants per m^2) to be released.
- A description of any precautions to maintain the genetically modified plant at a distance from sexually compatible plant species and to minimize or prevent pollen or seed dispersal.
- A description of the methods for post-release treatment of the site or sites. These are likely to include ploughing of the site, irrigation to encourage germination of any seed in the soil, the removal of any plants that sprout by spraying with an appropriate total herbicide and a period of one to two years when the site is kept fallow and monitored. The consequences of not doing this are evident from the "ProdiGene" incident described in Chapter 5.
- A description of post-release treatment methods for the genetically modified plant material.
- A description of monitoring plans and techniques.
- A description of emergency plans in the event that an undesirable effect occurs during the trial or that the plants spread.
- The likelihood of the genetically modified plant becoming more persistent than the recipient or parental plants in agricultural habitats or more invasive in natural habitats.
- Any selective advantage or disadvantage conferred to other sexually compatible plant species, which may result from genetic transfer from the genetically modified plant.
- The potential environmental impact of the interaction between the genetically modified plant and target organisms.

- Any possible environmental impact resulting from potential interactions with non-target organisms.
- The likelihood and consequences of theft of modified material from the trial, vandalism of the trial, movement of modified material from the trial site on field machinery, loss of modified material as a result of incident during transit, loss of viable, modified material as a result of inadequate sampling and laboratory processing procedures and regeneration from seed or vegetative tissue left in the trial area.

   Clearly, providing all of this information is a lengthy and possibly expensive business. Applicants also have to pay a fee of £5000, not insignificant in the context of research budgets, and wait at least ninety days for a decision (at least ninety days because the clock is stopped if the applicant is asked for further details until the information is provided). This delay can be a problem because, obviously, seed has to be sown at a particular time of year.

   In contrast, researchers in the US who want to undertake a field experiment with GM plants can submit a one-page application on-line and receive a decision the next day. That is not to say that the Americans have got it right. Over 30 000 field trials of GM plants have been performed in the US on species from soybean to walnut. The vast majority of these have caused no problems and the American authorities have made the application procedure much easier as a result. However, they are discovering that with some GM plant experiments, such as those involving plants that make pharmaceutical products, they have to be more careful (see Chapter 5). Nevertheless, the fact that field experiments are so much more difficult in the UK than the US does put researchers in the UK at a competitive disadvantage. Despite that, and the fact that no problems have arisen from the many field experiments of GM plants that have taken place in the UK, it appears likely that it will become more rather than less difficult to undertake field experiments in the UK in the future.

Safety of GM Foods

The safety of foods derived from GM crops is assessed by another committee, the Advisory Committee on Novel Foods and Processes (ACNFP). ACNFP was established in 1988 to advise the responsible authorities in the United Kingdom on any matters relating to novel foods and novel food processes. Currently it comprises fourteen academic members with expertise in allergenicity, genetics, immunology, microbiology, nutrition and food toxins, as well as a consumer representative and an ethicist. It comes under the jurisdiction of the Food Standards Agency (FSA).

The assessment made by the ACNFP essentially follows guidelines endorsed by the World Health Organization (WHO) and the same guidelines are followed by almost all regulatory authorities around the world. At the heart of the process is the concept of substantial equivalence. This concept is often attacked by anti-GM campaigners who claim that it merely means showing that a GM plant or food derived from it is roughly the same as its non-GM counterpart. In reality substantial equivalence involves a comprehensive biochemical and molecular comparison of a GM food and its conventional equivalent and a detailed analysis of any differences.

The fact is that very few foods consumed today have been subject to any toxicological studies, yet they are generally accepted as being safe to eat. The difficulties of applying traditional toxicological testing and risk assessment procedures to whole foods, GM or otherwise, makes it pretty well impossible to establish absolute safety. The aim of the substantial equivalence approach, therefore, is to consider whether the genetically modified food is as safe as its traditional counterpart, where such a counterpart exists.

The process begins with a comparison between the GM plant or food and its closest traditional counterpart in order to identify any intended and unintended differences. These differences then become the focus of the safety assessment and, if necessary, further investigation. Factors taken into account in the safety assessment include:

- The identity and source of novel genes (in particular is the source a well-characterized food source or is it entirely new to the food chain).
- Composition of the plant and/or food derived from it compared with its traditional counterpart.
- Effects of processing/cooking.
- The methods used to make the GM plant.
- The stability and potential for transfer of the novel gene or genes.
- The nature of the protein encoded by the novel gene or genes.
- Potential changes in function of novel genes and proteins.
- Potential toxicity of novel proteins.

- Potential allergenicity of novel proteins.
- Possible secondary effects resulting from expression of the novel gene or genes, for example by disruption of a gene in the host plant, knock-on effects on metabolic pathways and changes in the production of nutrients, anti-nutrients, toxins, allergens and physiologically active substances.
- Potential intake and dietary impact of the introduction of the genetically modified food.

The technology available to undertake these sorts of studies has moved on tremendously in the last few years. The activity of thousands, sometimes tens of thousands of genes can be determined in a single experiment. Similarly, the amounts of thousands of proteins and metabolites present in a plant can be measured. These techniques of transcriptomics, proteomics and metabolomics are still under development but will revolutionize safety testing. However, it is possible, likely even, that changes in gene expression, protein synthesis and metabolite profiles will be found in new varieties of crops produced by crossing, mutagenesis and genetic modification. The difficulty will be in identifying those changes that are significant and making rational decisions about them while anti-GM campaigners look for any opportunity to create a scare-story.

Animal Studies

The World Health Organization is skeptical about the usefulness of animal feeding studies in the safety assessment of GM plants and foods. This skepticism arose in part from experience gained in the testing of irradiated food in the early 1990s. Animal studies are undoubtedly useful in the safety assessment of individual compounds such as pesticides, pharmaceuticals, industrial chemicals and food additives. It is relatively simple to feed such compounds to animals at doses sometimes far higher than humans would be exposed to and to identify any potential adverse effects on health. Foods, on the other hand, are complex mixtures of compounds. Animals cannot be persuaded to eat orders of magnitude more than humans can, and feeding only one type of food to an animal usually reduces the nutritional value of the diet, causing adverse effects that are not related directly to the material itself. For these reasons, relating any effects on the welfare of an animal to a particular genetic modification can be extremely difficult. There is also an ethical question of whether it is right to undertake studies on animals if the results are unlikely to be meaningful.

Nevertheless, the WHO recommends animal studies if other information available on the GM plant or food is inadequate, particularly if the novel protein in the GM food has not been present in the food chain before. It also supports the use of animal testing of proteins produced by novel genes before the gene is used in biotechnology. In other words for the gene to be engineered into a micro-organism so that large amounts of its protein product can be made and purified. The individual protein can then be used in toxicology studies.

Allergenicity

I have already discussed the possible use of genetic modification to remove allergens from crop plants. While this is still in the early stages of development and is unlikely to be possible in all cases, it has great potential. Meanwhile, one of the major concerns expressed about the use of genetic modification in plant breeding is that it could lead to an increase in the number of food allergens. It could be argued, as with all aspects of safety, that genetic modification is no more or less likely to lead to an increase in food allergens than other methods in plant breeding. Nevertheless, the industry and regulators are clear that every precaution should be taken to ensure that no new allergens are introduced into the food chain through genetic modification.

In fact, the biotechnology industry is bullish about its ability to predict whether or not a GM food is likely to be more allergenic than its traditional counterpart and to detect new allergens before a product reaches the market if they get their predictions wrong. A typical procedure described by Astwood and co-workers (*Food Allergy*, 2nd Edition, Blackwell Science, 1996) combining molecular analyzes of the novel gene and protein, immunoassays and skin prick tests is shown in Fig. 4.1.

European Union Regulations

The EU recognizes two different types of field release of GM crops, one for research purposes only (a Part B release) and one for commercial release (a Part C release). Permission for a Part B release can be granted by an individual Member State. However, applications for a Part C release

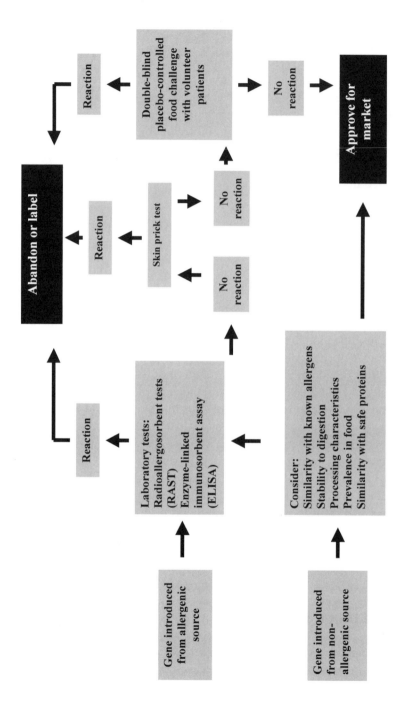

Fig. 4.1. Decision tree showing the assessment and testing of possible allergenicity in GM foods, as described by Astwood and co-workers (*Food Allergy*, 2nd Edition, Blackwell Science, 1996).

anywhere in the European Union have to be approved by the European Commission.

The way that the system is supposed to work is that applications are first submitted to one of the fifteen Member States. This state then becomes the lead Competent Authority (CA) for that application and has ninety days to complete its assessment. If the United Kingdom is the CA for an application the dossier is reviewed by the Joint Regulatory Authority, comprising the Department for the Environment, Food and Rural Affairs (DEFRA), the Scottish Executive, the National Assembly for Wales and the Department for the Environment in Northern Ireland. Advice is taken from ACRE, ACNFP and, if the crop is to be fed to animals, the Advisory Committee on Animal Feedingstuffs (ACAF). The application is considered to be at Stage 1A prior to review by ACRE and 1B when ACRE has given its advice to Ministers. Once the ninety days has passed the CA submits the application to the European Commission with an accept or reject recommendation. This is Stage 2.

Stage 3, if acceptance has been recommended, involves circulation of the dossier to the other fourteen Member States. The fourteen states have sixty days to make comment. Then if all Member States approve the marketing application, the lead CA issues what is known as a Part C marketing consent, which applies across all Member States.

If a Member State objects to the application, the Commission passes the dossier to its own Scientific Committee on Plants (SCP), a group of experts very similar to ACRE that considers exactly the same questions that ACRE considers. The SCP can consult two other committees, the Scientific Committee on Animal Nutrition (SCAN) and the Scientific Committee on Food (SCF), the equivalents of ACAF and ACNFP. This is Stage 4.

In Stage 5, if the SCP recommends that approval be granted, the Commission asks the Members States to vote again, this time by the Qualified Majority Voting (QMV) procedure. If there is a QMV in favor the lead Member State should issue consent. Otherwise, it moves on to Stage 6, in which the decision is referred to the Council of Ministers. Supposedly, the Council of Ministers can only reject the Commissions recommendation by a unanimous decision. Otherwise, in Stage 7, the Commission instructs the lead CA to grant consent. An individual Member State can still ban a particular GM crop in its own country, even if it has Part C consent.

All this to get a new crop variety on the market. Furthermore, in reality the Commission has had a *de facto* moratorium on the approval of new GM crops since 1998 because six member countries, France, Italy, Denmark, Greece, Austria and Luxembourg, block every application. Thirteen applications for consent to market a GM crop have been pending for the entire period since. Efforts to break this impasse at a meeting of the Environment Ministers of the Member States in October 2002 failed to reach agreement, despite the fact that the Commission has stated that its moratorium is probably illegal. The situation is a shambles that is damaging trade relations and the development of the biotechnology industry in Europe.

Those crops that were granted Part C consent before the moratorium began in 1998 are listed in Table 4.1. In addition, glyphosate-tolerant soybean, insect-resistant maize and slow-ripening tomatoes have consent for importation and use in human and animal food. The only significant commercial use of GM crops in Europe is in Spain, where insect-resistant maize has been grown successfully for several years.

A new Directive 2001/18/EC on the deliberate release of GM crops in Europe has just been implemented. It sets a ten year limit on the length of marketing consents, makes public consultation before a release compulsory (it was already compulsory in the United Kingdom) and sets out the requirement to have a post-release monitoring plan. The aim of post-release monitoring is to confirm any expected adverse effects of the GM plant and to identify unexpected effects on the environment or human health. Obviously, it will mean that GM crops and their products will have to be traceable through the food chain, potentially a huge task.

This new directive was supposed to lead to a relaxing of the moratorium on granting consents, but it has not. Furthermore, it risks making the marketing of GM crops so difficult that the biotechnology industry simply abandons Europe.

Labeling

The first GM plant product to come onto the market in the United Kingdom was paste made from slow-ripening tomatoes. The product was sold through two supermarket chains, Sainsbury and Safeway, clearly marked with a large label stating that it was made from GM tomatoes. The food retail industry intended to pursue this policy for all foods derived from GM plants, at least until consumers were familiar with the new technology. However, these plans were thrown into confusion in late 1996

Table 4.1. GM crops holding Part C consents for cultivation in the EU.

Company	Species	Trait
Amylogene	Potato	Modified starch
Aventis (now Bayer)	Oilseed rape	Herbicide tolerance
	Maize	Herbicide tolerance
	Oilseed rape	Herbicide tolerance
Bejo Zaden	Chicory	Herbicide tolerance
Monsanto	Maize	Insect resistance
	Maize	Herbicide tolerance
	Fodder beet	Herbicide tolerance
	Cotton	Insect resistance
	Cotton	Herbicide tolerance
Pioneer (part of DuPont)	Maize	Insect resistance and herbicide tolerance
	Maize	Insect resistance
Ciba Geigy/Novartis (now Syngenta)	Maize	Insect resistance and herbicide tolerance
Novartis (now Syngenta)	Maize	Insect resistance
Zeneca (now Syngenta)	Tomato	Delayed ripening

when shipments of that year's harvest of soybean and maize imported from the US arrived. Both contained at that time approximately 2% GM material with the GM and non-GM all mixed together.

For some reason, this appeared to take both the industry and authorities at national and European level by surprise. There was no legislation covering the labeling of GM foods and there was no agreement with the Americans to supply segregated GM and non-GM produce. In subsequent years representatives of the European food industry did negotiate with the Americans for non-GM soybeans in particular. The Americans could supply, at a premium price, soybeans from crops that had been grown from non-GM seed, and they could keep the non-GM produce segregated through the distribution chain. Nevertheless, they could not guarantee that there would be absolutely no mixing of GM and non-GM soybeans. The Europeans would not pay the asking price anyway.

Later, European food processors turned to Brazil for a supply of non-GM soybeans, although that supply is unlikely to last much longer given the penetration of herbicide-tolerant GM soybeans into Brazil. In the meantime, the food industry had a choice between labeling all products that contained the American soybean or maize and labeling none of it. They chose the latter. This turned out to be a mistake, because when the GM issue came into the public eye again a couple of years later consumers felt that GM food had been introduced behind their backs.

Labeling controls covering GM foods were finally introduced in Europe in 1997, with further updates in 1998 and 2000. The regulations require any food containing GM material to be labeled. They do not as yet cover processed food ingredients such as refined vegetable oils or sugar that are produced from a GM plant source but do not contain DNA or protein. This exception was made because it was felt that, since these products are indistinguishable from those produced from non-GM sources, it would be impossible to enforce the legislation (governments are always reluctant to introduce laws that they cannot police). New labeling laws will remove this exemption, despite the likelihood that it will lead to widespread fraud.

Foods that contain small amounts (below 1%) of GM material as a result of accidental mixing with a GM product are also exempt from the labeling laws. The EU regulations also exempt GM food sold in restaurants and other catering outlets. However, the United Kingdom extended the law to require caterers to provide written or verbal information covering GM foods to their customers. In practice, this usually takes the form of a cop-out statement in small print on a menu or pinned to a wall stating that GM ingredients are not used knowingly.

The aim of the GM food labeling laws is to enable consumers to make a choice, the argument being that however carefully the safety of GM foods has been assessed and tested, some people will choose not to eat them. Few would argue that giving consumers information is a bad thing, but inevitably the label is interpreted by consumers as a warning. Furthermore, there is a limited amount of space on a food label and it is questionable whether or not the fact that a product derives from a GM plant is the most important information to include. Nor is the law logical. A GM plant making its own insecticidal Bt protein but not sprayed with insecticide has to be labeled as GM while one on which the Bt insecticide has been sprayed can be labeled as organic (which consumers interpret wrongly as not having been sprayed with anything).

Despite this, the European Commission is seeking to extend GM labeling requirements to animal feed as well as human food and to include any food produced from a GM organism, regardless of the presence or absence of novel genetic material (in other words oils, sugar and other products that do not contain DNA or protein will no longer be exempt). Even this did not go far enough for some members of the European Parliament when it met to debate the issue, some calling for compulsory labeling of meat, milk and eggs obtained from animals fed on GM feed. So far this amendment has been rejected.

Strangely, while the labeling laws covering GM crops and their products are being tightened, there remains no legislation covering the labeling of foods such as most processed cheeses, yoghurts and many others that contain enzymes made in GM bacteria. This is absurd and our increasingly cynical trading partners are likely to smell protectionism on the part of the European Commission, since Europe produces only small amounts of GM crops while the use of GM enzymes in the European food industry is widespread.

The SCIMAC Agreement

In 2000 the UK government was faced with the prospect of the first GM crops becoming available for commercial cultivation in the United Kingdom while the public remained fearful and hostile. In an attempt to instil confidence and to satisfy the demand that the potential effects of GM crops on the environment should be tested more thoroughly before they were grown widely the government struck a deal with the biotechnology industry. This was called the SCIMAC (Supply Chain Initiative for Modified Agricultural Crops) agreement. The industry undertook not to commercialize GM crops before 2003. In the meantime, farm-scale trials would be carried out of the first generation of GM crops proposed for use in the UK in order to determine what effect they would have on the environment. The trials management committee includes representatives from English Nature, the government agency responsible for nature conservation in England, and the Royal Society for the Protection of Birds.

The government stated that commercialization after the farm-scale trials were completed would not proceed if the GM crops were shown to have an adverse effect on biodiversity. This has never before been asked of an innovation in agriculture and is still not asked of new non-GM varieties or of new or old farming methods.

The GM crops in the trials are herbicide-tolerant oilseed rape and fodder maize from Aventis (now owned by Bayer), and sugar and forage beet from Monsanto. However, in the United Kingdom, GM crops that are approved for commercial release still have to undergo standard new variety trials carried out by the National Institute of Agricultural Botany (NIAB). Only the maize variety, Chardon LL, has undergone these so far.

The first results of the farm-scale trials program will be published in the first half of 2003. It remains to be seen whether they will allay public fears. The anti-GM pressure groups who campaigned vociferously for more testing to be undertaken distanced themselves from the trials soon after they were announced and will no doubt dismiss the results if they do not suit them.

Safety Assessment and Labeling Requirements in the US

New GM crop varieties have to undergo field trials in the US in the same way that they do in Europe. However, much less detail is required in the risk assessment of the variety before the trial can go ahead. In fact, the procedure has been relaxed as more field trials have been undertaken and no problems have ensued. The total number of field trials of GM crops that have been run in the US now runs into the tens of thousands, covering a variety of crop species and traits. This may be reversed as applications are made to test crops that produce pharmaceuticals or have other non-food uses. The ProdiGene and StarLink episodes discussed in Chapter 5 suggest that they ought to be.

Test and commercial releases of GM crops in the US are controlled by the Animal and Plant Health Inspection Service (APHIS) within the United States Department of Agriculture (USDA). Commercial use of GM crops that produce their own insecticide, such as Bt, is controlled by the Environmental Protection Agency (EPA). The safety of foods derived from GM crops is assessed by the Food and Drug Administration (FDA). In fact, there is no legal requirement for companies to seek the approval of the FDA but all of them do. Companies ask the FDA for an Advisory Opinion either on a specific characteristic of their product or on its suitability as a food. Liability for problems that arise after release rests with the company that markets the crop.

The first GM plant product to come onto the market in the US was Calgene's Flavr Savr tomato (see Chapter 3). It underwent more than four years of comprehensive pre-market tests that were submitted to both the FDA and the USDA and were published for public comment. Calgene requested two separate Advisory Opinions from the FDA, one on the use of the kanamycin resistance marker gene, the other on the status of the Flavr Savr tomato as a food.

The FDA issued a preliminary report that all relevant safety questions about the Flavr Savr tomato had been resolved. This was ratified by its Food Advisory Committee in a public meeting in April 1994. The FDA announced its findings that the Flavr Savr tomato was as safe as tomatoes bred by conventional means in May 1994 and Calgene began to market the new product shortly after.

The labeling laws in the US are quite different from Europe. Products derived from GM plants do not have to be labeled as such. However, products that are significantly different from their conventional counterpart, such as high oleic acid soybean oil (see Chapter 3), do have to be labeled. In other words products are labeled according to their properties, not how they were made.

Published Test Results from Animal Feeding Experiments

Much is made by some of the fact that food safety test results are rarely published in science journals and therefore do not undergo the peer review process. Science journals are much more enthusiastic about publishing innovative and novel research than the results of essentially routine safety tests, which can make it difficult to find one that will publish such a study. It should also be remembered that test results are reviewed by committees of experts, a process that exceeds that of peer reviewing, and reports of the deliberations of the ACNFP are available to anyone who asks for them. Nevertheless, some studies have been published in the science press, notably those that have assessed the nutritional value of new varieties used for animal feed. Here is a summary of some of them:

- Safety assessment of 1-aminocyclopropane-1-carboxylic acid deaminase protein expressed in delayed ripening tomatoes, by Reed and co-workers, published in 1996 in the *Journal of Agricultural and Food Chemistry* (Vol. 44, pp. 388–494). The study analyzed GM tomato plants with delayed fruit ripening. The novel protein in the plants, an enzyme called 1-aminocyclopropane-1-carboxylic acid deaminase (ACCd), was tested under simulated mammalian digestive conditions and administered to mice at a dosage of up to 602 mg/kg of body weight (a 5000-fold safety factor relative to the average daily consumption of tomatoes). The mice showed no adverse effects.

- Evaluation of transgenic event 176 "Bt" corn in broiler chickens, by Brake and Vlachos, published in 1998 in *Poultry Science* (Vol. 77, pp. 648–653). This study compared broiler chickens fed with GM insect-resistant maize with chickens fed on non-GM maize. No differences were observed between the birds.

- Feeding value of corn silage estimated with sheep and dairy cows is not altered by genetic incorporation of Bt176 resistance to *Ostrinia nubilalis,* by Barriere and co-workers, published in 2001 in the *Journal of Dairy Science* (Vol. 84, pp. 1863–1871). This study evaluated insect-resistant GM maize and conventional maize in three separate feeding trials involving sheep and cattle. No differences were observed between the animals fed the GM diet and those fed the non-GM diet.

- Genetically modified feeds in animal nutrition first communication: *Bacillus thuringiensis* (Bt) corn in poultry, pig and ruminant nutrition, by Aulrich and co-workers, published in 2001 in *Archives of Animal Nutrition* (Vol. 54, pp. 183–195). The study again involved comparisons between an insect-resistant GM maize variety and its non-GM equivalent, this time fed to poultry, pigs and cattle. Again, no differences were found between animals on the GM diet and those on the non-GM diet.

- Genetically modified feeds in animal nutrition second communication: Glufosinate tolerant sugar beets (roots and silage) and maize grains for ruminants and pigs, by Bohme and co-workers, published in 2001 in the *Archives of Animal Nutrition* (Vol. 54, pp. 197–207). In this study pigs were fed GM gluphosinate-tolerant sugar beet and maize and compared with pigs fed an equivalent non-GM diet. No differences were found.

- The fate of forage plant DNA in farm animals: A collaborative case-study investigating cattle and chicken fed recombinant plant material, by Einspanier and co-workers, published in 2001 in *European Food Research and Technology* (Vol. 212, pp. 129–134). This study investigated the fate of ingested GM plant DNA in cattle and chickens being fed a diet containing conventional maize or GM insect-resistant maize. Gene fragments from the novel gene in the GM maize were not detected in any sample from either cattle or poultry.

- Nutritional assessment of feeds from genetically modified organisms, by Flackowsky and co-workers, published in 2001 in the *Journal of Animal and Feed Sciences* (Vol. 10, pp. 181–194). Digestion and feeding experiments were carried out with broilers (insect-resistant (Bt) maize), layers (insect-resistant maize, gluphosinate-tolerant maize), pigs (insect-resistant maize, gluphosinate-tolerant maize, gluphosinate-tolerant sugar beet and glyphosate-tolerant soybeans), sheep (insect-resistant maize silage, gluphosinate-tolerant maize silage) and cattle (insect-resistant maize silage). No differences were observed between GM-fed animals and animals fed on non-GM diets.

- Safety assessment of the neomycin phosphotransferase-II (NPTII) protein, by Fuchs and co-workers, published in 1993 in *Bio-Technology* (Vol. 11, pp. 1543–1547). The NPTII protein, which was present in Flavr Savr tomatoes and other GM plants, including potatoes, was shown to degrade rapidly under simulated mammalian digestive conditions. It caused no adverse effects when administered to mice at a dosage of 5000 mg/kg of body weight (a million-fold safety factor relative to the average daily consumption of potato or tomato).

- The feeding value of soybeans fed to rats, chickens, catfish and dairy cattle is not altered by genetic incorporation of glyphosate tolerance, by Hammond and co-workers, published in 1996 in the *Journal of Nutrition* (Vol. 126, pp 717–727). In this study, animal feeding experiments were conducted on rats, broiler chickens, catfish and dairy cows to compare glyphosate-tolerant soybean with its conventional equivalent. The feeding value of two glyphosate-tolerant varieties was found to be comparable to that of non-GM soybeans.

- Expression of the insecticidal bean α-amylase inhibitor transgene has minimal detrimental effect on the nutritional value of peas fed to rats at 30% of the diet, by Pusztai and co-workers, published in 1999 in the *Journal of Nutrition* (Vol. 129, pp. 1597–1603). Rats were fed GM peas and non-GM peas; the weight gain and tissue weights of rats fed either of the two diets were not significantly different from each other. (See Chapter 5 for a discussion of another of Pusztai's experiments with GM food.)

- Glyphosate-tolerant corn: The composition and feeding value of grain from glyphosate-tolerant corn is equivalent to that of conventional corn, by Sidhu and co-workers, published in 2000 in the *Journal of Agricultural and Food Chemistry* (Vol. 48, pp. 2305–2312). Glyphosate-tolerant GM maize was evaluated in a poultry feeding study. Results from the poultry feeding study showed that there were no differences in growth between chickens fed with the GM grain and those fed with a non-GM equivalent.

None of these studies was long-term (no food, GM or otherwise, has been subjected to a long-term study of its effects on health). It should also be remembered that the World Health Organization and other expert bodies believe that their assessments based on the demonstration of substantial equivalence between GM plants and their traditional counterparts and investigations into any differences gives a much better indication of risk or lack of it.

5 ISSUES THAT HAVE ARISEN IN THE GM CROP AND FOOD DEBATE

Farmers will only grow genetically modified crops if their products are accepted by consumers. While this does not appear to be a problem in many parts of the world, consumers in Western Europe are split over the risks and advantages of GM crops. The proportion of consumers strongly opposed to or fearful of GM crops is declining, but slowly, and the food processing and retail industries still perceive there to be market advantage in presenting themselves as GM-free.

Exactly how "GM-free" the European food industry really is could be open to question. There was undoubtedly some concern within the US soybean industry in 1999 that they would lose market share in Europe (the maize industry might have been expected to be equally concerned but they export very little maize to Europe anyway because of trade tariffs imposed by the EU). However, the American Soybean Association reported in 2000 that sales of unsegregated (i.e., GM) soybeans to the EU had increased, not decreased, that year. There had been enquiries from EU customers regarding purchase of identity-preserved, non-GM soybeans, but there was a reluctance to commit to specific quantities or to pay the premium price for it. Clearly, Europeans and/or their animals are eating American GM soybeans in one form or another.

Many UK food processors claim to be buying non-GM soybeans from Brazil. In fact, there is widespread growing of GM soybeans in Brazil, although this does not yet have official approval (GM soybeans are approved for import and food use in Brazil, but not for cultivation). UK food processors recognize this and there are some signs that they realize that the present situation cannot go on much longer.

In the meantime, however, great damage has been done to the UK plant biotechnology industry. Syngenta, Monsanto, Aventis (now Bayer), DuPont and Unilever have all closed or down-sized their plant biotechnology operations in the UK in the last few years. Researchers have also been put at a great disadvantage. If a scientist in the US wishes to undertake a field trial of a genetically modified crop he/she can submit a one-page application on-line and receive approval the next day. In the United Kingdom a scientist has to submit a complicated and detailed risk assessment, pay a fee of £5000 (a lot of money in the context of research budgets) and wait for at least ninety days before receiving approval (Chapter 4).

Exactly why European consumers have been so much more fearful of GM crops than other consumers is not clear. A recent poll showed that 66% of consumers in China, Thailand and the Philippines believed that they would benefit personally from food biotechnology during the next five years. A different poll in the US found that 71% of US consumers would be likely to choose produce that had been enhanced through biotechnology to require fewer pesticide applications. Polls in the UK and Europe continue to show much less favorable attitudes amongst consumers.

Part of the answer lies in the reluctance of Europeans to trust their governments or scientific experts. GM foods were launched in Europe shortly after the epidemic of bovine spongiform encephalopathy (BSE) in the UK cattle herd had lead to one of the biggest food scares in UK history. Rightly or wrongly, consumers felt that they had been given the wrong advice by scientists and government ministers on the safety of beef. However, food "scares" are not unique to the UK and Europe.

Another reason for consumer antipathy towards GM crops in Europe is that the debate has been dominated by anti-GM pressure groups. European consumers have been bombarded with inaccurate information, half-truths and wild scare-stories. Even if they do not believe the more hysterical of these stories, why should they take the risk of buying GM food products? In the late 1990s much of the popular media in the United Kingdom joined the anti-GM bandwagon. Perhaps the nadir of the campaign came on 26 April 1999 when *The Daily Mail* carried the headline "Scientists Warn of GM Crops Link to Meningitis" on its front page. The article continued, "The nightmare possibility of GM food experiments producing untreatable killer diseases has been underscored by senior government scientists". I never discovered who these "senior government scientists" were. The story was retracted the next day in a short statement on an inside page.

There is not a lot that scientists can do in the face of a campaign that is waged at that level. In any case, few scientists are trained to deal with the media or the public and most find it a difficult and uncomfortable experience. The "correct" answer to the question "Are GM foods safe" is the one that I gave in Chapter 4, "There is an international consensus amongst experts in the field that GM is not inherently more risky than other methods in plant breeding; GM foods are examined under a rigorous assessment system that goes beyond that applied to other foods; the safety of GM food has been considered by the World Health Organization, the European Union, the OECD and many national governments and none has concluded that there is any evidence of adverse effects on health". However, there is little chance of getting through the first line of this statement before a television interviewer interrupts or the audience starts thinking about something else. The opposing argument that GM foods are unsafe or untested may not be true but it is much simpler to put across.

So what are the issues and concerns that have been raised during the GM crop debate? The rest of this chapter outlines and addresses some of them.

Are GM Foods Safe?

No scientist will ever describe anything as completely safe. However, if we accept that new crop varieties produced by non-GM methods are safe enough, it is reasonable to judge the safety of GM crops in comparison. The major arable and horticultural crops grown and consumed in western Europe have been developed over centuries or even millennia. Consequently, they are assumed by the consumer to be safe and wholesome. However, most, if not all, of these crops contain compounds that are potentially toxic or allergenic. In most cases, these compounds have probably evolved to provide protection against animal predators or pathogenic micro-organisms and it is therefore not surprising that they are also toxic to humans. They include glycoalkaloids in potatoes, cyanogenic glycosides in linseed, glucosinolates in *Brassica* oilseeds and proteinase inhibitors in soybean and other legume seeds.

It is very doubtful whether these or many other generally accepted foods would be approved for food use were the toxins introduced by genetic modification. Nevertheless, with some exceptions, the introduction of new types and varieties of food crops produced by conventional breeding requires no specific testing for the presence of allergens and toxins, even if genes have been introduced from exotic varieties or related wild species.

The products of genes introduced by genetic modification are readily identified in a GM plant. They may also be isolated in a pure form, either from the species of origin, from the GM plant or after expression in a micro-organism. The pure protein can then be tested in detail and its presence in processed foods monitored. In contrast, it is virtually impossible to predict or characterize all of the changes in food composition that may result from conventional plant breeding.

On top of all this, as I have described in Chapter 4, GM crops and foods are examined under a rigorous assessment system. Consequently, it would not be unreasonable to argue that GM foods may be safer than food derived from non-GM varieties.

Will Genetic Modification Produce New Food Allergens?

It has been suggested that consumption of GM foods could lead to increases in allergenicity. In fact, there has already been a widely reported case of increased allergenicity in a GM plant line, but the problem was detected and the product never reached the market. A methionine-rich 2S albumin storage protein gene from Brazil nut was expressed in soybean in order to increase the content of methionine, an amino acid, in soybeans used for animal feed. The protein was subsequently shown to be an allergen, as are a number of related 2S albumins from other species, and the program was discontinued.

This case certainly illustrates the need for caution. However, the fact that the problem was identified before commercial material was produced demonstrates the high level of awareness of such problems in the plant biotechnology industry and the effectiveness of the "in house" screening programs described in Chapter 4. The biotechnology industry takes the view that release of new allergenic products into the food chain is entirely avoidable.

Is it Ethical to "Tinker" with the DNA of Plants?

This question has been raised famously by Prince Charles. He has been quoted as saying that "Mixing genetic material from species that cannot breed naturally, takes us into areas that should be left to God". Of course, plant breeders were mixing genetic material from sexually incompatible species long before genetic modification was developed (see the section in Chapter 1 on triticale as an example). Nevertheless, this statement encapsulates the views of a significant number of people. It is a pity that the point could not have been made without the use of overblown language, such as the term "Frankenstein Foods".

The philosophical counter argument to this view is that all life on Earth has a common ancestor if you go back far enough and that the species alive today are part of an evolutionary continuum, not separate entities. So genes encoding proteins involved in cellular control mechanisms that evolved long ago, such as the control of cell division, are closely related and instantly recognizable to those that work on them, whether they come from yeast, invertebrates, plants or man. So how can it be fundamentally wrong to move genes between species?

The pragmatic counter argument is that it is not where a gene comes from or how it is brought into a plant breeding program that is important but what it does. A gene producing a poison that is produced inadvertently by radiation mutagenesis or crossed into a crop plant from a wild relative is clearly much more dangerous than a gene producing a benign protein that is introduced into a crop plant by genetic modification. Similarly, genes imparting herbicide tolerance can "flow" into weed plants by cross-pollination regardless of whether they have been engineered or crossed into a crop plant or produced by radiation mutagenesis.

Part of the reaction against moving genes between species is an emotional one. The use of an "antifreeze" gene from a fish to engineer frost resistance into tomatoes, for example, caused a much greater reaction than the use of a bacterial gene to engineer insect resistance into maize, despite the fact that the tomatoes were not a commercial product whereas the maize was. So far biotechnology companies have avoided using animal genes in GM crops for food use but are using them to make pharmaceutical products in GM plants. Clearly, the industry believes that consumers have different attitudes when they are buying and using pharmaceuticals to when they are buying and consuming food.

As with many technologies, the question of ethics cuts both ways. World population doubled in the second half of the last century. A food crisis was only averted by the green revolution, which involved a combination of new, higher-yielding crop varieties and the use of agrochemicals. According to the US Census Bureau and the United Nations Department of Economic and Social Affairs, the annual growth in world population is slowing. In the 1960s world population growth averaged just over 2% per year. By the 1990s it had declined to 1.4% per year and the forecast for 2005 is less than 1.2%. Nevertheless, world population is predicted to continue to increase throughout this century. Furthermore, the decline in population growth is linked to increased income, which itself is linked with a demand for a better diet, particularly a demand and ability to pay for more meat. Meat production requires a lot more land than crop production.

Clearly, the need for agriculture to continue to produce more and better food is not going to go away. Yet agriculture faces some severe challenges, such as the effects of climate change, limited water supply, soil erosion, salination and pollution. At the same time, the amount of wild land that can be taken into agricultural use is running out and many people would like to conserve what remains of it. The people who will suffer if plant breeders and farmers fail to deliver the necessary improvements are those who are already short of food. Too much has been made of the idea that GM crops will "feed the world"; that is a burden that no new technology can carry. But is it ethical for the well-fed people of western Europe to block any technology that might contribute to improved crop yield and quality?

An example of the impact on world agriculture and food supply of western Europe's reluctance to accept GM food is the recent controversy over the supply of food aid to southern Africa. The famine-threatened countries Malawi, Swaziland, Lesotho, Zimbabwe and Mozambique have finally agreed to accept food aid from the US that contains GM maize. Unfortunately Zambia, despite the dire need of its people for food, continues to reject the aid.

This issue arose in late summer 2002 and rumbled on for several months. At first, encouraged by European activists who warned of supposed health risks associated with GM food in a bid to turn the US food aid into another anti-GM food campaign issue, all of these countries rejected the food. As it became clear that people were going to die without the food aid (and only the US produces enough surplus to meet the demand) the activists became silent, fearing a backlash. As the bogus health issue receded the problem became one of future trade with Europe. The African governments feared that some of the imported seed would mix with seed being planted for future harvests and this would jeopardize trade with Europe once the drought was over.

At first, the EU refused to reassure the Africans that GM crops were acceptable, afraid that this would threaten their trade bans on the importation of some GM foods from the US. However, it eventually relented and gave the reassurances that were asked for. Even so, this was not sufficient to persuade Zambia to accept the food aid, and at present its citizens are at unnecessary risk of starvation as a result.

GM Crops "Do Not Work"

A long-running strategy of anti-GM campaigners is to claim that GM crops are flawed in some way. In other words there is a problem inherent in the technology that means that no GM crop will deliver what it promises. GM crops have been linked by activists with crop failures in India and Indonesia, neither of which have grown GM crops commercially yet, and with farm failures in the US, where farmers continue to use GM crops in increasing numbers.

In its recent report, "Seeds of Doubt", the Soil Association, a UK organic farming group, claimed that the use of GM crops in the US had reduced profitability, reduced yields and raised costs through increased herbicide use. Herbicide-tolerant soybean was cited as an example of a crop that had failed to meet expectations of higher yields. I have referred already (Chapter 3) to the much more extensive report prepared by the National Center for Food and Agricultural Policy. That report cites herbicide-tolerant soybeans as the most successful GM crop in improving farm profitability. Increased profits are gained through reduced costs, not improved yield. Indeed, improved yield was never a target for herbicide-tolerant crops in a developed country where farmers have access to a variety of weed control measures. So why do the two reports differ? Perhaps because the Soil Association report, as it states itself, did not cover positive experiences.

Not all new crop varieties, GM or otherwise, are successful. Indeed, one of the first GM varieties on the market, the "Flavr Savr" tomato, has already come and gone. However, the success of GM crops overall has been almost unbelievable. Ten years ago, few people even in the biotechnology industry would have predicted that by 2002 more than half of the world's soybean crop would be GM. The major GM crop varieties have now been available in the US for six years, long enough for farmers to make a judgement. The National Agricultural Statistics Service of the United States Department of Agriculture reports that GM maize acreage in the US rose from 26% in 2001 to 34% in 2002, GM soybean acreage rose from 68% in 2001 to 75% in 2002 and GM cotton acreage from 69% in 2001 to 71% in 2002. This could only happen if farmers liked the new varieties and it is absurd to pretend otherwise.

Greenpeace used the argument that GM is a quick fix that will not work in an attempt to debunk Golden Rice (see Chapter 3). In their press release of February 2001 they described Golden Rice as "fool's gold", claiming that an adult would have to eat at least 3.7 kg of dry rice (twelve times the normal intake of 300 g) to get the daily recommended amount of pro-vitamin A. The press release coincided with the start of a breeding program undertaken by the International Rice Research Institute in the Philippines to cross Golden Rice with local rice varieties.

Ingo Potrykus, the developer of Golden Rice, replied that the calculations of Greenpeace were based on recommended daily allowance values for western consumers, who have access to as much vitamin A as they care to eat. The levels of pro-vitamin A achieved with Golden Rice are in the 20–40% range of the recommended daily allowance in a 300 g serving. Nutritional experts involved in the project believe that this level should have a significant effect in preventing blindness and other symptoms associated with severe vitamin A deficiency. Even so, the full recommended daily allowance would be provided by 0.75 to 1.5 kg dry weight, not 3.7 kg.

How effective the Golden Rice technology turns out to be in the end will depend on the levels of pro-vitamin A that can be obtained when the trait is crossed into local rice varieties. Countries such as the Philippines and India that potentially have much to gain from it are assessing it carefully in relation to other possible solutions to vitamin A deficiency. What is not refuted by anyone is that despite all other efforts to prevent it, vitamin A deficiency causes blindness in 500 000 children every year and millions of deaths. The fact that anti-GM groups clearly hope that Golden Rice does not provide the answer simply because it is the product of genetic modification does them no credit.

Did Tryptophan Produced by Genetic Modification Kill People?

Tryptophan is an amino acid and is essential to all living things in order for them to make proteins. Plants, fungi and bacteria make their own but animals, including Man, do not and need to acquire it in their diet. Experiments with rats fed tryptophan-deficient diets suggest that tryptophan deficiency delays growth, development and maturation of the central nervous system. As well as its role as an amino acid required for protein synthesis, tryptophan is a precursor for a neurotransmitter, serotonin. There is no real evidence that tryptophan deficiency is a problem in developed countries since it is plentiful in meat and dairy products (vegetarian diets may be low in it) but health supplement manufacturers claim an extensive range of benefits associated with taking tryptophan as a supplement.

Many pharmaceutical companies around the world manufacture tryptophan, usually from selected strains of bacteria. One of these, Showa Denko in Japan, used a GM strain of *Bacillus amyloliquefaciens* to produce the raw product in the late 1980s. In 1989 the company changed the strain of bacterium that it was using and reduced the number of purification processes that it used to clean up the raw product. The result was that certainly hundreds, probably thousands of people in the United States who bought the product under various brand names became ill with eosinophilia myalgia syndrome (EMS). Thirty-eight people died. Tryptophan was withdrawn as an over-the-counter product in the US and the Food and Drug Administration has never allowed its return, although it can be obtained on prescription.

An investigation into the incident found that several contaminants that should have been removed in the purification process were present in the finished product. Why they were still present is not clear since Showa Denko did not release the results of its own internal inquiry. It is also still not certain which contaminant was responsible for the illnesses and deaths. No manufacturer is currently using GM bacteria to make tryptophan, but problems with contaminants and the resulting condition of EMS still occur sporadically, so the problem is clearly one of old-fashioned chemical engineering, not genetic modification. Perhaps the solution is to eat a good steak instead of using an expensive health supplement.

The irony of this story for plant biotechnologists is that it concerns GM bacteria, yet it is used as an argument against the genetic modification of plants. Meanwhile, the use of GM bacteria to produce medicines, vaccines, food additives and supplements steams ahead.

The Monarch Butterfly

In 1999 a study conducted by John Losey and his team at Cornell University found that caterpillars of the monarch butterfly that were forced under laboratory conditions to eat large quantities of pollen from GM insect-resistant maize suffered higher mortality levels than caterpillars that were not fed the pollen. The study was published in the journal *Nature* and caused an international furore. Indeed, because of the iconic status of the monarch butterfly in the US, there were fears that the study might cause the sort of public backlash against GM crops there that had occurred in Europe. As it turned out, although the issue was more grist to the mill for the media and anti-GM activists in Europe, Americans remained broadly supportive of plant biotechnology.

The *Cry1A*, or Bt gene (see Chapter 3) that is engineered into maize to make it insect-resistant is toxic to caterpillars. The gene is designed to be active everywhere in the plant, including the pollen (subsequent varieties may contain genes that are designed only to be active where they are needed), so it is no surprise that Monarch butterfly caterpillars that ate the pollen did not thrive. However, this was a laboratory experiment. Monarch butterfly larvae eat milkweed, not maize pollen (in the experiment the pollen was spooned onto milkweed leaves so that the larvae had not choice but to eat it). Even as the first report was published, experts were extremely skeptical that maize pollen would ever accumulate in such amounts on milkweed in the wild. Field-based studies published subsequently bore this out.

Similar laboratory-based experiments have shown that the survival rate of predator species such as lacewings and ladybirds can be reduced if they are fed exclusively on prey species that are feeding on GM insect-resistant plants. None of these results have been replicated in the field. It should be remembered that spraying caterpillars, lacewings, ladybirds and other insects with pesticide, which is the practice for non-GM maize, kills them all outright. The use of GM insect-resistant plants in the field can lead to an increase in beneficial insects. With GM maize, for example, the farmer does not have to use an early spray against the corn borer. The farmer may need to spray against other pests but, if this can be avoided, benign predatory insects can thrive. If this happens, the predatory insects prevent a late infestation of red spider mites, reducing pesticide use even further.

Butterfly populations are affected by many things of course. Nevertheless, since GM crops have been grown widely in North America since 1996 it is worth asking how the Monarch butterfly has done in that time. Mike Brannon of Associated Press wrote an article in October 1999 entitled "Monarch Butterfly Population on the Rise across America". In it, Donald Lewis, Professor of Insect Studies at Iowa State University was quoted as saying "It's been a good year for butterflies in general, and we're looking forward to a prolonged Fall so we can enjoy the show." Not much evidence of a negative impact of GM crops there. Unfortunately, unusually cold weather in Mexico where the Monarch butterfly over-winters had a serious effect on the species' numbers in 2001–2002.

As soon as the initial report was published in 1999 and the controversy erupted, Losey was at pains to point out that the study was a preliminary one and that field studies would be required to find out the real impact of GM crops on butterfly populations. This raised eyebrows amongst the science community because *Nature* does not normally publish preliminary reports. What is more, why did the journal not require the researchers to compare the effects of force-feeding caterpillars on GM maize pollen with the effects of pesticide spraying? There is little doubt that the study would not have been considered worthy of publication without the GM crop tag and that publication was rushed through in order to generate headlines and publicity.

The Pusztai Incident

The other incident that brought the GM issue to the fore in 1999 was a press release made by a scientist, Dr. Árpád Pusztai, who was working at the Rowett Institute in Scotland. Dr. Pusztai had produced GM potatoes that were engineered to produce a type of lectin. Many plants produce lectins as natural insecticides but most lectins are unsuitable for use in plant biotechnology because they are poisonous to animals, including humans. The lectin gene that Pusztai used came from snowdrop. When the GM potatoes were fed to rats, the rats suffered from lectin poisoning, which is, perhaps, not very surprising. However, Pusztai fed another group of rats with ordinary potatoes that had been spiked with the lectin.

These rats, too, suffered from lectin poisoning but in the view of Pusztai they did not suffer as badly as those fed with the GM potatoes. Pusztai deduced that the process of genetic modification had somehow made the potatoes more poisonous than they would have been otherwise.

The normal method of dissemination of scientific results is through the scientific press. Every article is reviewed by at least two independent experts in the field, the aim being to make sure that experiments have been conducted properly and that the deductions made by the authors of the article are justified by their data. Pusztai decided to bypass this system and called a press conference before submitting his results for peer review. The story was picked up by the general media and the affair turned into a circus.

The affair became such an important public issue that the experiments and results were reviewed by a group of scientists appointed by the Royal Society, who concluded that the work was "flawed in many aspects of design, execution and analysis" and that "no conclusions should be drawn from it". Not a particularly edifying end to a previously distinguished career.

Pusztai's work may have been dismissed by his peers but the damage had already been done. Pusztai retired but became a hero of anti-GM campaigners, to whom he still acts as a consultant.

"Superweeds"

The notion that the use of GM herbicide-tolerant crops could lead to the creation of "superweeds" (uncontrollable weed plants) is a favorite of anti-GM activists. The story goes that a gene that imparts herbicide tolerance in a GM crop could "flow" into related wild species through cross-pollination and the resulting "superweeds" would be uncontrollable.

As I have described in Chapter 3, herbicide tolerance is not a novelty that came along with genetic modification. All crop plants are naturally tolerant of some herbicides and the genes that impart that tolerance are just as likely to "flow" into populations of wild relatives as genes introduced by genetic modification. Furthermore, broad-range herbicide tolerant non-GM varieties of maize and soybean are now on the market in competition with GM varieties. They escape the regulations applied to GM crops; a ridiculous situation in which crop varieties are controlled according to the method used to make them rather than their characteristics.

Unlike non-GM crops, the potential environmental impact of cross-pollination of GM crops with wild species has to be assessed case by case, taking into account the species and genes involved. Maize and potato, for example, do not cross with any wild species in the United Kingdom (although forced crosses can be made between potato and black nightshade in the laboratory) and wheat does not cross with any native plant species to produce fertile hybrids. Oilseed rape will cross with other cultivated and wild *Brassicas*, including Chinese cabbage, Brussels sprouts, Indian mustard, hoary mustard, wild radish and charlock. The extent of such crossings in agricultural systems is the subject of continuing research, but it does not necessarily mean that GM oilseed rape represents a threat. Recent studies carried out in Canada, where there are millions of acres of GM herbicide-tolerant oilseed rape, have not found any hybrids between the GM crop and wild relatives.

Even if a hybrid between a GM crop plant and a wild relative did arise, it would only prosper if the gene involved could confer a competitive advantage. Herbicides do not exist outside agriculture, so herbicide-tolerant genes would be unlikely to persist in the wild. Weed populations that acquire herbicide tolerance genes from crop plants or that evolve tolerance themselves could become a problem for farmers and if the problem were to become too serious the herbicide and the GM crop that went with it would become useless. Farmers would have to turn to other varieties and different herbicides. Farmers have been aware of these problems since herbicides became used widely, long before genetic modification was developed.

Farmers will also have to be careful that GM crops with different herbicide-tolerant traits do not cross to produce multi-tolerant hybrids. Such hybrids have arisen between gluphosinate- and glyphosate-tolerant varieties of oilseed rape in Canada and have caused problems for farmers who have expected to be able to clear their fields with one of the two herbicides but have found that some oilseed rape plants unexpectedly survived the treatment.

Superbugs

Another emotive term used by anti-GM campaigners is "superbugs". Ironically, this idea is something that they latched on to after a warning issued by the Environmental Protection Agency (EPA) of the US. The EPA had been given the responsibility of monitoring and controlling GM plants that had been engineered to be insect-resistant through the introduction of a *Cry* gene (see Chapter 3) because it was already responsible for controlling the use of the Bt pesticide. The EPA argued that since both the GM plants and the pesticide used the same insecticidal protein the two had to be controlled by the same agency. This makes sense in some ways but other GM crops are controlled by the Animal and Plant Health Information Service (APHIS) within the US Department of Agriculture and this splitting of responsibility for GM crops may have contributed to the StarLink fiasco (see below).

A long-term concern for the EPA with the Bt pesticide has been the risk of emergence of resistant insects ("superbugs" in the parlance of anti-GM activists). This concern pre-dates the development of GM crops but was heightened in 1996–1997 when it became clear that the use of GM crops containing the *CryIA* gene was likely to become widespread very rapidly. The EPA's answer was to insist that farmers using GM crops containing the *Cry1A* gene would have to plant a proportion of non-GM crop as well. This would provide a refuge in which insects that had developed resistance to the effects of the Bt protein would not be at a selective advantage (in fact they would be at a selective disadvantage). According to the EPA's predictions of how the insect population would respond the proportion of non-GM crop would have to vary according to what other insect-resistant GM crops were being grown in a particular area.

If the *Cry1A* gene were to "flow" into weed plants through cross-pollination the models on which the EPA bases its predictions of how insect populations are likely to respond would break down. For this reason the EPA proposed banning GM insect-resistant cotton in areas of the US where cotton has wild relatives.

As is always likely in the US, this proposed interference from a Federal Government Agency in Washington in what varieties farmers could or could not grow went down very badly in the Mid-West. In response to criticism from farmers' representatives, the EPA published a report that stated that if its recommendations were not followed there would be widespread resistance to Bt in insect populations within three to five years. Both the insect-resistant GM crops and Bt pesticides would then be useless. The EPA got its way and the refuge policy was introduced. So far it appears to have been very successful. However, activists in Europe lifted the worst-case scenario prediction out of the EPA's report and used it as one of the justifications of their opposition to GM crops.

Segregation of GM and Non-GM Crops: Coexistence of GM and Organic Farming

Unless GM food is accepted universally, which seems unlikely in the foreseeable future in Europe, it is important that alternatives remain available to allow consumers to exercise choice. For imported food-stuffs, European suppliers will either have to buy produce from countries where GM crops are not grown, which will become increasingly difficult, or pay farmers overseas to grow non-GM varieties. This means paying a guaranteed price to a farmer to use old-fashioned varieties and high chemical inputs. As I have stated already, European buyers have so far been reluctant to commit to paying an increased price for non-GM soybean from the US.

For crops grown in the UK, the main issue will be segregation of GM and non-GM crops and food. Segregation could break down through accidental mixing of GM and non-GM seed for planting, by cross-pollination between GM and non-GM crops (not a problem for inbreeding species such as wheat) or by mixing of the product between the farm gate and the consumer. Some inadvertent mixing is almost inevitable and the production of certifiably GM-free food is therefore likely to be expensive if not impossible.

The organic farming industry in Europe and the US has attempted to corner the GM-free market. Indeed, presenting itself as GM-free has been possibly its most successful marketing strategy ever, at least in Europe. As part of this strategy the industry persuaded authorities in the US and Europe to write GM-free into the official definition of organic produce. This has lead to the bizarre situation where an organic farmer can spray Bt pesticide (not to mention some much more toxic pesticides) onto a crop and still sell it as organic, but cannot grow a crop that has been engineered to produce the Bt protein itself. This situation cannot last forever. As consumers become more familiar with GM foods and more GM products with consumer benefits come onto the market, there will eventually be a demand from consumers for some GM products to be grown organically.

In the meantime, the organic industry in North America has made a rod for its own back. As GM crops become more and more popular with other farmers it is becoming increasingly difficult for organic farmers to guarantee that their produce is GM-free. In Europe, the organic industry uses this as an argument for GM crops to be banned. Since organic farmers represent a tiny minority of the total this would be absurd, but that does not mean that it will not happen.

Clearly, if organic and GM farming is to co-exist there will have to be tolerance on both sides. Organic farmers will have to accept that there might be some cross-pollination of their crops from GM crops planted in adjacent fields. On the other hand, farmers who have paid for expensive GM seed in order to produce a high-value product will have to accept that some of the seed produced by the crop will result from cross-pollination from old-fashioned varieties grown by the organic farmer next door. Recent studies in Australia suggest that a tolerance level of 1% (i.e., a crop with 1% GM content could still qualify as organic) would be sustainable even for out-breeding crops such as oilseed rape.

Antibiotic Resistance Marker Genes

Another topic that has generated a great deal of debate, some of it wildly overblown, is the use of antibiotic resistance genes as selectable markers. The use of marker genes to select cells that have been modified with genes of interest is discussed in Chapter 2, and antibiotic resistance genes have been extremely valuable in the development of GM technology. Many scientific bodies around the world, including the World Health Organization and regulatory committees set up by the European Union and several national governments have considered the safety of antibiotic genes in food and have concluded that those that are being used do not represent a health threat. The British Medical Association, however, have expressed reservations, although the scientific arguments on which they based their conclusions were flawed, and the United Kingdom's Advisory Committee on Novel Foods and Processes has called for the development of alternative marker systems. The biotechnology industry considers antibiotic resistance marker genes to be safe but a public relations liability and it is unlikely that they will be used in commercial GM varieties in the future. Nevertheless they will continue to be used in basic research.

The supposed risk associated with the use of antibiotic resistance marker genes is that they will somehow find their way into gut or soil bacteria and from there into disease-causing bacteria. In fact, the antibiotic resistance marker genes used in plant biotechnology confer resistance to antibiotics that are not used at all in oral medicines, such as kanamycin and neomycin. Furthermore, these genes are designed to work in the plant and would not be active in bacteria.

Antibiotic resistance marker genes are also used in the genetic modification of bacteria, including bacteria used for the cloning, manipulation and bulking up of genes to be used in plant transformation. Where whole plasmids are used to transform plant cells by particle bombardment (see Chapter 2), such a gene may be incorporated into the plant genome along with the gene of interest. The antibiotic resistance marker genes used for the genetic modification of bacteria include one that imparts resistance to ampicillin, which is used in medicine. This gene was reported to be present in an early variety of GM insect-resistant maize.

So is there any risk associated with the presence of such a gene in a GM plant? Firstly, the gene is integrated into the genome of the GM plant and the risk of horizontal gene transfer from plant genomes to soil or gut bacteria under natural conditions is extremely low. Indeed, I am not aware of any evidence that horizontal gene transfer can occur between genes in plant genomes and soil or gut bacteria. Furthermore, where did these genes originate from in the first place? The ampicillin resistance gene came from a human gut bacterium, *Escherichia coli*. In fact, even a cursory examination of the information available shows that many natural bacterial species contain an ampicillin resistance gene, including *Kluyvera ascorbata*, *Pyrococcus furiosus*, *Proteus mirabilis*, *Bacillus subtilis*, *Klebsiella pneumoniae* strain H18, *Klebsiella pneumoniae* strain G122, *Pseudomonas aeruginosa*, *Pyrococcus furiosus*, *Staphylococcus aureus*, *Synechocystis* sp. PCC 6803, *Sinorhizobium meliloti*, *Yersinia pestis* strain CO92, *Mycobacterium leprae*, *Deinococcus radiodurans*, *Vibrio cholerae* and many soil bacteria. In other words the "scare" over antibiotic resistance genes used in plant biotechnology is that they could somehow find their way into bacterial populations that already have them.

Antibiotic resistance marker genes are only maintained in bacterial populations if they impart a selective advantage. Antibiotic resistant strains of pathogenic bacteria do represent a health threat, but they arise naturally and thrive because of the sloppy management of antibiotics in human and animal medicine, not because of the use of antibiotic resistance marker genes in biotechnology.

Patenting

Patenting law was devised for two reasons, firstly to allow inventors to benefit from their inventions, secondly to bring inventions into the public domain so that they would not die with their inventors. The invention must be described in sufficient detail to allow others to repeat it and must satisfy three criteria for the patent to be approved: it must be novel, it must be useful and it must be "non-obvious" even to an expert in the field.

All of this seems quite laudable and the system has generally achieved its objectives. Patents protect inventions for up to twenty years, enabling inventors to prevent competitors from copying their invention or to license use of the technology at a price. Any new gadget is likely to be covered by a plethora of patents and licensing agreements.

Traditionally, patents were not awarded to new plant varieties or to strategies devised by plant breeders to improve crops. Indeed, until relatively recently plant breeders had almost no commercial protection at all for their new varieties. This changed in 1961 with the advent of Plant Breeders' Rights, a form of intellectual property designed specifically to protect new varieties of plants. Plant Breeder's Rights were drawn up at The International Convention for the Protection of New Varieties of Plants (the UPOV Convention) in Geneva in 1961 and became law in the United Kingdom in 1964. They were revised and strengthened in 1991, although the new regulations did not come into force in the UK until 1997.

Plant Breeder's Rights enable the holder of the rights to prevent anyone from producing, reproducing, offering for sale or other marketing, exporting, importing, conditioning for propagation or stocking a new variety without a license from the holder. However, Plant Breeder's Rights do not prevent another breeder from using a variety in their own breeding program.

Since genetic modification was developed, many patents have been filed covering the use of specific genes, types of genes, gene promoters, GM plant varieties and various technologies used to transform plants. Most of these patents have never been tested in court and it is still unclear to what extent they will be effective. Many people in the industry are skeptical about their usefulness but continue to file patents if only to ensure that they are not shut out by someone else doing so.

More confusion is generated by the fact that different countries have different patenting laws and interpretations. European patent examiners have been very stringent in making sure that biotechnology patents satisfy the three criteria listed above while their American counterparts have been more inclined to accept a patent filing and let the courts sort it out later if the patent is challenged. Canada does not allow patents that cover plants or animals at all. The World Trade Organization is attempting to uphold patents covering plant varieties worldwide under the General Agreement on Tariffs and Trade (GATT) but it has its work cut out.

The advantage for a company in holding a patent on the use of a particular gene in plant biotechnology is that the gene can then only be used under license. Anyone who wants to incorporate the gene into their own breeding programs or even farmers who want to grow a crop containing the gene can only do so with permission from the company holding the patent. The biotechnology company has the right to charge a royalty for use of the gene or of crops that contain the gene. It is therefore able to make money out of its technology in ways that Plant Breeder's Rights would not allow.

Patenting law does not allow for so-called "bio-piracy", the raiding of a developing country's biological resources by a Western company. A patent on a particular species of tree, or a traditional crop, would not satisfy the three criteria listed above and would be thrown out. However, it is possible to patent the idea of using a particular gene from any source to engineer a trait into a crop plant. Scientists in countries such as India and China are well aware of this and are keen to exploit the huge biological potential of their native flora, which dwarfs that available to scientists in temperate countries.

Loss of Biodiversity

It is widely believed that the introduction of GM crops will reduce biodiversity in agriculture, or more accurately it will reduce the genetic diversity available to plant breeders in the form of different crops and varieties. The great success of, for example, herbicide-tolerant soybean might appear to support this idea. However, the fact that a particular trait becomes popular with farmers does not necessarily mean that a single variety will be grown by everyone. The glyphosate-tolerant trait has been licensed by Monsanto to many other plant breeders who have crossed the original line with their own varieties to produce new varieties that carry the herbicide-tolerance gene but are suited for particular local growing conditions. Approximately one hundred and fifty different seed companies now offer glyphosate-tolerant varieties of soybeans.

Similarly, the original Golden Rice is unlikely to be grown commercially anywhere. It is already being crossed into breeding lines in the Philippines and India in order to produce locally-adapted varieties carrying the high vitamin A and iron content traits.

Plant breeders are acutely aware of the risks of too much inbreeding and the potential of wild and unusual genotypes for providing genes that might improve a crop breeding line. For this reason, they conserve old and wild varieties in case they can be used in a breeding program. Potato breeders, for example, have access to over 200 cultivated potato lines and 8000 wild potato relatives that have been characterized. This is more genetic diversity than they are ever likely to know what to do with.

The Dominance of Multinational Companies

Much of the basic science and technology that underpins plant biotechnology was developed in the public sector in the US, Europe and Japan and plant biotechnology continues to attract considerable government funding for research all around the world. The commercial development of GM crops, however, is dominated by six multinational companies. These are BASF, Bayer (who recently acquired Aventis), Dow, DuPont, Monsanto and Syngenta. The domination of any industry by a small number of companies is always a concern, but it is hardly unique to plant biotechnology. Ironically, although the multinationals have retreated from the UK they will be able to return later if public opinion changes. The small and medium sized home-grown companies and start-ups that might have competed with them are unlikely to survive.

The StarLink and ProdiGene Fiascos

StarLink is a trade name given to several GM maize varieties produced by Aventis (now part of Bayer) that, like those varieties discussed in Chapter 3, have been engineered to be resistant to insects by the insertion of a *Cry* gene from *Bacillus thuringiensis*. However, unlike the successful GM maize varieties grown for human consumption, StarLink contains the *Cry9C* version of the gene instead of the *Cry1A* version. It also contains the gene imparting tolerance of the herbicide gluphosinate.

Problems arose with StarLink because, unlike the product of the *Cry1A* gene, that of the *Cry9C* gene does not break down easily in the human digestive system and is relatively heat resistant. For this reason, StarLink has never been approved for human consumption. However, in 1998 the Environmental Protection Agency (EPA) of the US approved StarLink for commercial growing as an animal feed. It is not clear to what extent the other federal agencies involved in GM crop assessment, the Food and Drug Administration (FDA) and the Animal and Plant Health Information Service (APHIS) were consulted on this decision. The EPA set a zero-tolerance level for the use of StarLink in human food.

Maize is an outbreeder and some cross-pollination between StarLink and maize varieties destined for human consumption was inevitable. This should undoubtedly have been foreseen by the EPA and Aventis but apparently it was not. In mid-September 2000, the *Washington Post* reported that StarLink maize had been detected in processed maize snack foods already on the shelf in grocery stores. The products were recalled, the FDA became involved and in the end three hundred maize products had to be taken off the shelf. Products were also recalled in Japan and Korea.

Aventis instructed its seed distributors in the United States to stop sales of StarLink seed corn for planting in 2001 and voluntarily cancelled its EPA registration for StarLink. However, Aventis predicted that it would take four years for StarLink to clear the food chain entirely. The company agreed to buy back the entire StarLink crop of 2000 at a premium price, something that must have cost them upwards of US$100 million. The 1999 crop was already beyond recall. The issue is likely to rumble on for several years as farmers, distributors and processors seek compensation.

Another incident involving the accidental mixing of a non-food GM crop with food intended for human use occurred in November 2002, although in this case the crop in question did not go beyond the farm. The biotechnology company involved was ProdiGene, a company that specializes in developing GM crops to produce pharmaceutical products. Clearly, these crops are not meant to enter the human food chain.

In 2001, ProdiGene contracted a Nebraska farmer to grow one of its experimental GM maize varieties. However, after the crop had been harvested the land was not treated any differently to land that had had an ordinary crop growing on it. The next year the farmer planted soybean destined for human consumption on the same land. Inevitably a small number of maize plants arising from spilled seed grew amongst the soybeans and a tiny amount (65 g) of corn stalks were discovered in the harvested soybean seed. The Food and Drug Administration ordered the soybean crop, worth US$2.7 million, to be destroyed.

This was almost unbelievably sloppy on the part of the company and/or farmer involved. Such a release in the UK would have to be monitored regularly during all stages of development. Post harvest the plants and/or seed would be stored in a secure area; all ventilation would be enclosed and all dust from extraction units or from drying and storage areas would be inactivated by autoclaving (heating under pressure). All straw and similar material would be returned to the release site and burnt prior to cultivation into the soil. After harvest, the site would be ploughed and monitored. Irrigation would be used to encourage germination of volunteers and any re-growth or volunteers would be removed by spraying with an appropriate total herbicide. No crop would be sown on the site at least for the following year, probably the one after as well. The release site would be monitored for two years and any re-growth or volunteers would be removed by spraying with an appropriate herbicide or by cultivation.

All this would apply to a crop designed eventually for food use, never mind one making a pharmaceutical product. Clearly, similar practices should have been used in this case. However, the case does not change the fact that the use of GM plants to produce pharmaceutical products has enormous potential.

The StarLink and ProdiGene incidents should not have been allowed to happen and they highlight the need to segregate crops designed for food and non-food uses. Regulatory authorities will have to learn from these incidents and decide whether outbreeding species such as maize are suitable plants for the purpose of making pharmaceutical and other non-food products. In fact, earlier in 2002 the United States Department of Agriculture ordered a different ProdiGene GM maize trial in Iowa to be destroyed to prevent cross-pollination with food maize growing nearby. Decisions such as that should be made before the crop is planted and some crops for which there is zero tolerance of presence in the food chain will have to be contained under glass. Nevertheless, it should be remembered that this problem is not unique to GM crops. In the United Kingdom and elsewhere, for example, there is long experience of growing non-GM oilseed rape varieties to produce industrial oils. These oils are not suitable for human consumption and there is a strict tolerance level enforced for the non-food rapeseed in the food rapeseed harvest.

The Cauliflower Mosaic Virus 35S RNA Gene Promoter

One of the barmier scare-stories put about by anti-GM campaigners is that the use of the cauliflower mosaic virus 35S RNA gene promoter (usually called the CaMV 35S promoter, see Chapter 2) in plant biotechnology represents some sort of human health risk. Cauliflower mosaic virus infects cauliflowers (hence its name) and other plants in the *Brassica* (cabbage) family. It does not infect animals, including humans, at all, but it infests most of the *Brassica* crop of the United Kingdom to some extent, and probably always has done. It is not, therefore, a good candidate for a food scare.

The CaMV 35S gene promoter is only a small part of the viral genome. It is used in plant biotechnology where a gene introduced by genetic modification is required to be active everywhere in the plant. The supposed risk associated with it is that some sort of recombination could occur between this promoter and an animal virus, producing a new "supervirus". It is not clear exactly how this is supposed to happen, why it would be more likely to occur when the CaMV 35S promoter is integrated into a plant genome as opposed to a viral genome, or why, even if it did occur, the resulting virus would be imbued with super powers.

Implications for Developing Countries

The effect that GM technology will have on developing countries is a question that has worried many people and there was certainly concern in developing countries either that they would become reliant on first-world companies for their seed or, conversely, that they would be denied access to GM technology altogether. These worries still persist.

Developing countries are not all the same, of course; there is a big difference between countries such as India and China that have a large science base of their own and poorer countries that do not. Both India and China are investing heavily in plant biotechnology.

It is important to note the potential benefits of GM technology for developing countries, as well as the problems. The potential of high vitamin GM crops such as Golden Rice, virus resistant and high yielding crops, and crops making edible vaccines to relieve famine, malnutrition and disease in developing countries is obvious. One of the more satisfying developments in the GM crop story of recent years is that developing countries are making their own decisions on the matter, much to the chagrin of Western anti-GM pressure groups.

An example of a scientist from a developing country who advocates strongly that the potential of GM crops should at least be explored is Florence Wambugu, the director of the African Center of the International Service for the Acquisition of Agri-biotechnology Applications (ISAAA), in Kenya.   Thanks to the influence of Wambugu and others of like persuasion, Kenya has moved further and faster than most sub-Saharan African countries in using plant biotechnology. Wambugu, who studied for her PhD at the University of Bath, has stated that "having missed the Green Revolution, African countries know they cannot afford to pass up another opportunity to stimulate overall economic development through developing their agriculture".

Wambugu is behind the development of GM sweet potato that is resistant to feathery mottle virus (Chapter 3). GM insect-resistant (Bt) maize varieties are also being field-tested; Kenya currently loses approximately 40% of its maize harvest to the stem borer.

"Terminator" Technology

The term "terminator" has been used to describe a method of using genetic modification to produce crop plants with sterile seeds. It has been used by anti-GM campaigners to show how the biotechnology industry will use GM to control agriculture. It is ironic, therefore, that the technology was developed in a United States Department of Agriculture research station, not in a private company. Furthermore, the idea was patented but so far as I know has never been taken any further. In fact, the issue demonstrates one of the advantages of the patent system: bringing ideas and new technologies into the public domain. It is not clear where the term "terminator" came from, since it is not used in the patent description. It is likely the invention of anti-GM campaigners. The "terminator" patent was acquired by the Delta and Pine Land Company and subsequently by Monsanto. However, Monsanto have never shown any inclination to use it.

It has been suggested that biotechnology companies could use "terminator" to force farmers to buy seed from them every year. Many farmers in the developed world do this anyway, in some cases because the variety that they use is a hybrid that would decline in performance in subsequent generations, or because using fresh seed prevents virus and other diseases passing from one year's crop to the next. The second reason applies to third world farmers just as much, but they rarely have access to fresh seed and have to save seed from one year to the next whether they like it or not. Nevertheless, in order to make farmers buy "terminator" seed, a company would have to link the "terminator" trait with a trait that farmers really wanted. Since "terminator" technology has never been used in a commercial variety, however, this is all hypothetical.

Conclusions

Genetic modification is now an established technique in plant breeding in many parts of the world. While not being a panacea, it does hold the promise of enabling plant breeders to improve crop plants in ways that they would not be able to through other methods in plant breeding. GM crops now represent approximately 6% of world agriculture, and are being used in developed and developing countries. Farmers who use them report one or more of greater convenience, greater flexibility, simpler crop rotation, reduced spending on agrochemicals, greater yields or higher prices and increased profitability at the farm gate as the benefits.

That does not mean that every farmer who has the option to use GM varieties does so. GM varieties enter a very competitive marketplace and suppliers offering alternatives reduce their prices or offer other incentives to make their own products competitive. Nor is it for me to tell farmers what they should or should not grow. What I would like to see is UK and European farmers given the choice of growing GM crops that have come through the testing and regulatory processes. Otherwise they will find it increasingly difficult to compete with overseas farmers just at a time when there is international pressure to dismantle and withdraw protective trade barriers and subsidies.

In the United Kingdom there is currently a voluntary moratorium on the growing of GM crops brought about by the SCIMAC (**S**upply **C**hain **I**nitiative for **M**odified **A**gricultural **C**rops) agreement (Chapter 3). This was an agreement between the agricultural biotechnology industry and the UK government in which the industry undertook not to commercialize GM crops until farm-scale trials had been completed. The first results of the farmscale trials will be published in 2003 and the UK government will have to make a decision to allow commercial planting to go ahead (assuming that farmers will want to grow the GM varieties), or try to persuade the industry to extend the moratorium or bring in legislation to ban GM crops altogether.

The delay in allowing plant biotechnology to develop in Europe has already damaged the European plant biotechnology industry significantly and is putting European agriculture at an increasing competitive disadvantage. Europe desperately needs politicians at all levels to show leadership on the issue but there is little indication that they will. Environment ministers of the European Union member countries meeting in October 2002 failed to agree on the lifting of the EU's *de facto* moratorium on new approvals of GM crops, despite the fact that the European Commission has stated that the moratorium is probably illegal. The fact is that powerful, multinational pressure groups continue to call the shots on GM crops and food in Europe and these groups remain implacably opposed to the use of the technology.

Public debate of a new technology is not a bad thing, of course, but genetic modification of plants is no longer a new technology. Eventually the debating has to stop, conclusions have to be drawn and decisions made.

Index